AF564823

A Colour Handbook on Rainfed Kharif Crops

A Colour Handbook on Rainfed Kharif Crops

Protection, Constraints and Mitigation Strategies

Authors

Reena
Sonika Jamwal
Anil Kumar
A.P. Singh
Vikas Abrol
R. Puniya
Jai Kumar

NEW INDIA PUBLISHING AGENCY

New Delhi – 110 034

NEW INDIA PUBLISHING AGENCY
101, Vikas Surya Plaza, CU Block, LSC Market
Pitam Pura, New Delhi 110 034, India
Phone: + 91 (11) 27 34 17 17 Fax: + 91 (11) 27 34 16 16
Email: info@nipabooks.com
Web: www.nipabooks.com

Feedback at feedbacks@nipabooks.com

ISBN: 978-93-85516-32-0

Composed, Designed and Printed in India

Foreword

As with abiotic causes of crop losses, especially the lack or excess of water in the growth season, extreme temperatures, high or low irradiance and nutrient supply, biotic stresses too substantially reduces crop production. Estimated losses due to pests in India is 33%, 26%, 20%, 6% and 6-8% by weeds, insects, diseases, rodents and other pests respectively. These account for crop losses up to 1980, 1300, 1000 and 300 crores of rupees, respectively caused by weeds, insects, diseases, rodents and other pests. As per the data available with GOI (2010), the value of loss caused by insect pests alone in million is Rs. 29450, 26100, 43551 and 41368 in maize, rapeseed-mustard, pulses and wheat respectively. These organisms may be controlled by applying physical (cultivation, mechanical, use of traps, *etc.*), cultural (crop rotation, intercropping *etc.*) biological (parasitoids, pathogens, predators *etc.*) and chemical measures (pesticides). Protection makes only 5-8% of the total cost during production in one vegetation. On the other hand, damages resulting from absence of proper care or poor performance can be huge. If protection is omitted or done superficially and inadequately, losses will be even higher.

In agriculture weeds not only cause huge reduction in crop yields but also increase cost of cultivation, reduce input efficiency, interfere with agricultural operations, impair quality, act as alternate hosts for several insect pests, diseases and nematodes. In non-crop areas, they affect aesthetic look of the ecosystem as well as native biodiversity, health and quality of life. Weeds are a big constraint in crop production and are responsible for heavy yield losses in almost all the crops. Losses due to weeds in major rainfed crops are 6-35% in wheat, 30% in rapeseed-mustard, 30-50% in maize and 17-40% in sesamum. A recent study undertaken at DWR-Jabalpur suggests that proper weed management technologies if adopted can result in an additional production of 103 million tonnes of food grains, 15 million tonnes of pulses, 10 million tonnes of oilseeds and 52 million tonnes of commercial crops, per annum, which in few cases are even equivalent to the existing annual production. This may amount to an additional income of Rs. 1,05,036 crores per annum.

Proper identification of the pest species; insect pests, diseases or weeds in field is essential before the implementation of suitable management options available.

A ready reference for proper identification of different species is a major bottleneck that is faced not only by researchers but also by field functionaries and farmers. In this regard, an attempt has been made by ACRA staff, Rakh Dhiansar, Bari Brahmana, SKUAST-Jammu to come up with a book on field problems of rainfed kharif crops containing the description, damaging symptoms and management strategies to be adopted, thereof along with their coloured illustrations of insects pests, diseases and weeds of major rainfed kharif crops. This publication shall be of immense use to UG and PG students along with other stakeholders interested in exact identification and management of pests.

I congratulate the authors for their painstaking efforts in bringing out this publication at an appropriate time.

Dr. Jag Paul Sharma
Director Research
SKUAST-Jammu

Date: 01-10-2017
Place: Jammu

Preface

Both abiotic and biotic factors are responsible for crop losses. Without preventive protection with pesticides, natural enemies, host plant resistance and other nonchemical controls, 70% of crops could have been lost to pests. Weeds produced the highest potential loss (30%), with animal pests and pathogens being less important (losses of 23 and 17%). Crop protection techniques have been developed for the prevention and control of crop losses due to pests in the field. Pests reduce crop productivity in various ways, classified by their impacts; stand reducers (damping-off pathogens), photosynthetic rate reducers (fungi, bacteria, viruses), leaf senescence accelerators (pathogens), light stealers (weeds, some pathogens), assimilate sappers (nematodes, pathogens, sucking arthropods), and tissue consumers (chewing animals, necrotrophic pathogens) etc. The efficacy of control of pathogens and animal pests only reaches 32 and 39%, respectively, compared to almost 74% for weed control.

Kharif season crops of rainfed areas are inflicted by several important insect pests, diseases and weeds. Correct identification of insect pests, diseases and weeds is therefore necessary not only for strict quarantine to check the spread of new pest species, but to achieve the desired productivity levels. The purpose of this publication entitled "A Colour Handbook on: Rainfed Kharif Crops Protection Constraints and Mitigation Strategies" is to assist the students, field researchers, scholars and farmers in correct identification of these pests and their management thereof. In this book, efforts have been made to describe the damaging symptoms, identification characteristics, biology and management techniques of more than 100 pest species. Coloured illustrations have been provided in this book for amateurs. An old Chinese proverb "one picture is worth of ten thousand words" is especially applicable to the identification.

For effectively managing these pest problems in the field, one needs to correctly identify them, know their biology, so as to hit the weakest point in their life cycle and then go for the appropriate management strategies to be applied, looking to the availability of resources. In this context the team of ACRA scientists, Dhiansar, SKUAST-Jammu have compiled the information on more than 100 pests with colour photographs, along with their description, damaging symptoms, biology and management techniques to be followed, which has practical importance and application in field. We hope this book would be of great help to students, researchers, quarantine officers, scientists and policy makers and would also serve as a ready reference for identification and management strategies of insect pests, diseases, weeds and nutritional deficiencies of rainfed kharif crop plants.

We thank the supporting staff of ACRA-Dhiansar, Bari Brahmana, SKUAST-J, who have helped us in coming up with this publication. The cooperation extended by Dr. J.P. Sharma, Director Research of the University is duly acknowledged. Thanks are also due to other staff members for their helping hand in successful completion of this book.

December, 2017 **Authors**

Contents

Section B: Major Diseases of Rainfed Kharif Crops

Section C: Major Weeds of Rainfed Kharif Crops

Section D: Nutritional Deficiency Disorders

Section - A
Major Insect Pests of Rainfed Kharif Crops

1

Cereals

MAIZE

In India, more than 130 insect species have been recorded to cause damage to maize crop.

1. Maize stem borer, *Chilo partellus* (Swinhoe) (Lepidoptera: Pyralidae)

Distribution

This is the most destructive pest of maize and sorghum in Sri Lanka, Pakistan, Afganistan, Uganda, Central and East Africa.

Damaging symptoms

Caterpillars damage by boring into the stem, cobs or ears. Young larvae first feed on leaves, making few shot holes and then bore downwards through the central whorl. Early attack (20-30 days after sowing) causes dead-heart.

Life cycle

Flat, oval, yellowish eggs are laid in clusters of upto 20 eggs. A single female lays approximately 300 eggs on the underside of the leaves, which hatches in 4-5 days. Grown up caterpillars are 20-25 mm long, dirty grayish while with black head and four brownish longitudinal stripes on the back. Adults are yellowish-grey moths, 25 mm across the wings. The larva undergoes five moults in a period of 14-28 days. It breeds actively during March-April to October and hibernates as full grown larvae in stubble, stalks, cobs, etc. during rest of the year. There are five generations in a year and the total life cycle is of three weeks.

Management Strategies

- Destruction of stubble, weeds and other alternate hosts by ploughing the field after harvest.

- Removal and destruction of dead hearts and early pin-hole damage.
- Destruction of crop residues.
- Installation of pheromone traps @ 5 per ha for monitoring and @ 20 per ha for mass trapping of male moths.
- Release of *Trichogramma* spp. when egg masses are initially noticed.
- Release of *Apanteles* sp. or *Microbracon* sp. adults synchronizing with the larval development.
- Apply Carbofuran 3G granules to the whorl of the plant @ 0.5 – 1.25 kg/ha per application through a bottle with a few hole in its cap. It should be applied 2-3 weeks after sowing or when borer injury on leaves is noticed.
- Use *Bacillus thuringiensis* (*Bt*) to minimise harm by all lepidopteran pests.
- Alternatively, spray with any of the insecticide Cypermethrin 10 EC or Fenvalerate 20 EC or Deltamethrin 2.8 EC @ 100 ml / ha or Carbaryl 50 WP @ 250 WP @ 250 g/ha.
- Thiamethoxam 12.6% + Lambda cyhalothrin 9.5%ZC (a combination product) may also be applied

Shot hole symptoms caused as a result of feeding by *C. partellus* larvae

Adult moths trapped in pheromone traps

C. partellus larva

C. partellus pupa

Life cycle or *Chilo partellus*

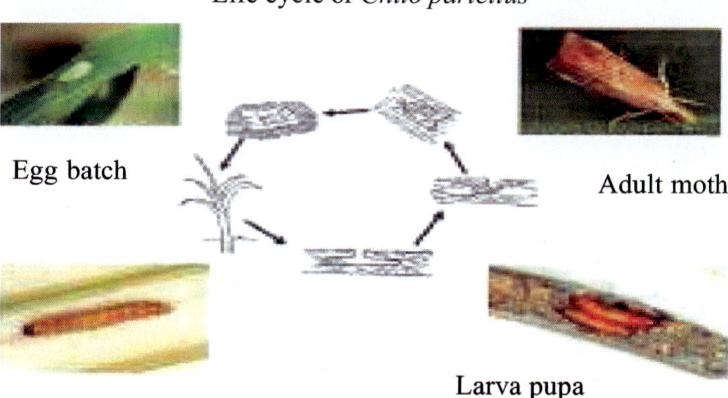

2. Asian maize borer, *Ostrinia furnacalis* (Guenee) (Lepidoptera: Pyralidae)

Distribution

It is widely distributed in SriLanka, China, Taiwan, parts of India, Bangladesh, Indonesia, Japan, Korea, Veitnam, Laos, Cambodia, Malaysia, Myanmar, Thailand and the Phillipines.

Damaging symptoms

The larvae make tunnel into the stalks and the ears. Maize plants are attacked 3-4 weeks after sowing. Larvae first feed on leaf tissue, then tunnel into the midrib and then feed on the tassels or bore into stalk and ears. When it attacks the base of the stalk, the plant collapses, dries up and dies.

Life cycle

Female moth is yellow to light brown, 27 mm across the wing span, while the male is dark brown with tapering abdomen. A single female lays upto 500 – 1500 eggs in its life time. The incubation period is of 3days and there are five larval instars. The larval period is of 27 days and the pupal period is of 6 days.

Management Strategies

Same as in case of maize stem borer.

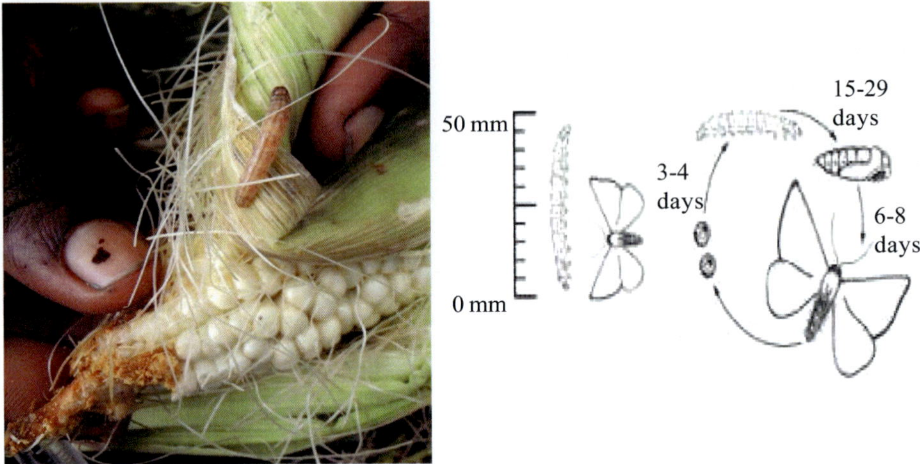

Ostrinia furnacalis larva   *O. furnacalis* Life cycle

3. Blister beetle, *Mylabris pustulata* Thunberg (Coleoptera: Meloidae)

Damaging symptoms

It is a polyphagous pest attacking flowers or inflorescence of several crop plants of Families Malvaceae, Leguminosae, Convulvulaceae, Graminae and Cucurbitaceae. The adult beetles attack the tassel and silk of maize cobs and devour them completely, resulting in great reduction in grain yield. Damage is more at higher altitude regions.

Life cycle

The beetle is quite large, approximately 3 cm in length and has six alternating bright orange and black bands against the dark background of body. These brightly coloured insects secrete compound containing Cantharidin. During August – September their population reaches its peak. The female lays about 100 – 2000 eggs in soil. Grubs on hatching, feed on soil dwelling insects. The first instar grub is called tringulin as it has three-clawed legs. Later instars are less active and they pupate in soil itself.

Management Strategies

- Handpick and destroy the adult beetles (wear gloves while handling them).
- The smoke of burned dead beetles deters the other live beetles in the field and they move to the nearby fields.
- Spray the crop with Carbaryl 50 WP @ 2 ml/L of water. Synthetic pyrethroids may be used for a quick knock-down effect.

Mylabris pustulata adult feeding on flower of okra and silk of maize

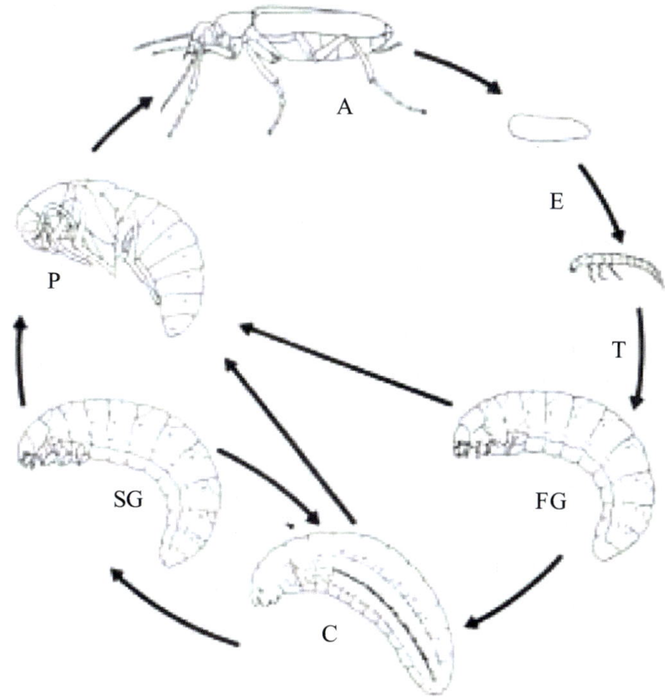

M. pustulata life cycle

A = adult, E = egg, T = first instar or triungulin, FG = first grub phase, C = coarctate phase in instar six or seven, SG = second grub phase, P = pupa.

SORGHUM

It is attacked by more than 150 insect species.

1. Sorghum midge, *Contarinia sorghicola* (Diptera: Cecidomyiidae)

Distribution

It was first reported from south India, now it has spread to all parts of India.

Damaging symptoms

Flowers appear damaged, earheads are devoid of grains as maggots feed on ovaries. In case of serious attack, earheads may appear blighted.

Life cycle

It is a tiny 2 mm, pinkish, fragile, mosquito like insect with bright orange abdomen and two transparent wings. Adults are short lived (males – few hours, females – 1-2 days). Females oviposit (30-100 eggs) on glumes of open or closed flowers. Maggots after hatching in 1-2 days, start feeding on ovaries. It undergoes pupation in 9-11 days within the damaged flowers. After three days they become adult.

Management strategy

- Both early and late maturing varieties should not be sown in the same area, as they provide continuous availability of flowers.
- The panicle and post harvest trash should be collected and burnt immediately to avoid carryover of the pest.
- These insecticides may be applied first at 90% panicle emergence and second after 7-10 days. – Carbaryl 50 WP or Malathion 50 EC.

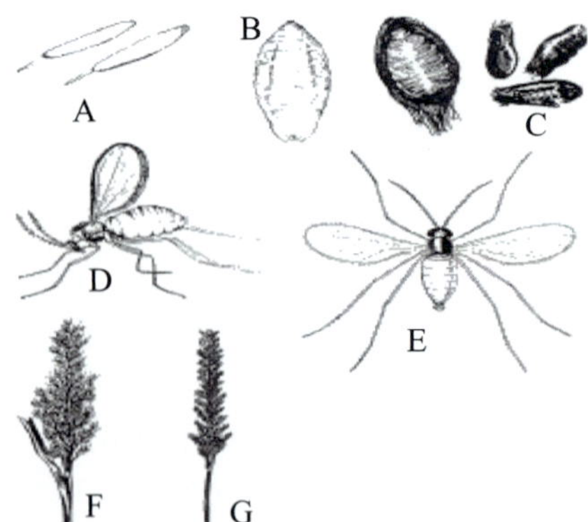

Sorghum midge. A. Eggs, B. larva, C. Pupa in cocoons, D. Adult female. E, Adult male. F. Normal sorghum head. G, Damaged sorghum

Contarinia sorghicola adult and life cycle

2. Sorghum ear head bug, *Calocoris angustatus* (Hemiptera: Miridae)

Distribution

This is the most destructive pest of sorghum in Southern India. It feeds on number of cereals, millets, grasses, etc.

Damaging symptoms

Both nymphs and adults feed on the green earheads, as a result of which, the grains remain chaffy or shriveled. In case of heavy infestation, the whole ear may become blackened at first and eventually dry up, producing no grains.

Life cycle

The bug lays cigar shaped eggs (150-200) under the glumes or in between anthers of florets. The eggs hatch in 5-7 days. There are 5 nymphal instars, which in three weeks time become adults. Adult is a small bug slender, greenish yellow, 5-8 mm in length, 1 mm in width.

Management strategy

- Spray 625 ml Malathion 50 EC or 250 ml Dimethoate 30 EC in 500 L water /ha.

Calocoris angustatus adult, egg and pupa

3. Sorghum shoot bug, *Peregrinus maidis* (Hemiptera: Delphacidae)

Distribution

This pest is serious on sorghum, millets and maize in South India.

Damaging symptoms

Both adults and nymphs suck the cell sap from leaf whorls, leaf sheaths as a result of which, leaves become yellow and plant growth is retarded. When the attack is severe, ears fail to emerge.

Life cycle

The female bug lays white, elongated, cylindrical eggs in groups of 1-4, in the midrib of leaves by making slits with the help of ovipositor. A single female lays upto 90-100 eggs in its lifetime. After hatching in 7-10 days, the nymphs moults five times to become adults in 16-18 days. The total life cycle is completed in 3–4 weeks.

Management strategy

- Same as ear head bug.

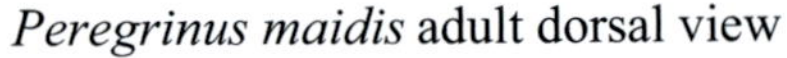

Peregrinus maidis adult dorsal view

P. maidis adult Side view

4. Sorghum shoot fly / stem fly, *Atherigona soccata* (Diptera: Anthomyiidae)

Distribution

It is widely distributed in Europe, Africa and Asia. It also attacks other cereals, maize, wheat, small millets, grasses, etc.

Damaging symptoms

It attacks at the six leaf stage of the crop. When the crop becomes 40-45 days old, it is seldom attacked by this pest. The tiny maggots on hatching, reaches in between the sheath and main shoot and bore into the stem. As maggots feed on the main shoot, growing point is destroyed, thus showing typical dead-heart symptoms.

Life cycle

Female oviposits on the underside of leaves. Eggs are flattened, elongate and somewhat boat shaped, provided with two wing like lateral projections. A single female lays upto 40 eggs in its lifetime. After hatching in 2-3 days, the maggots undergo pupation within 10-12 days. They pupate either in the soil or in the stem.

Management strategy

- Seed coating with Isofenphos 5 G @ 30g / 100 kg seed.
- Early sowing (10-15 days after the onset of monsoon).
- Apply Carbofuran 3G, Chlorpyriphos 5 G in furrows before sowing @ 3g per meter row length.
- Destroy infested plants 3-4 weeks after sowing.

Atherigona soccata adult

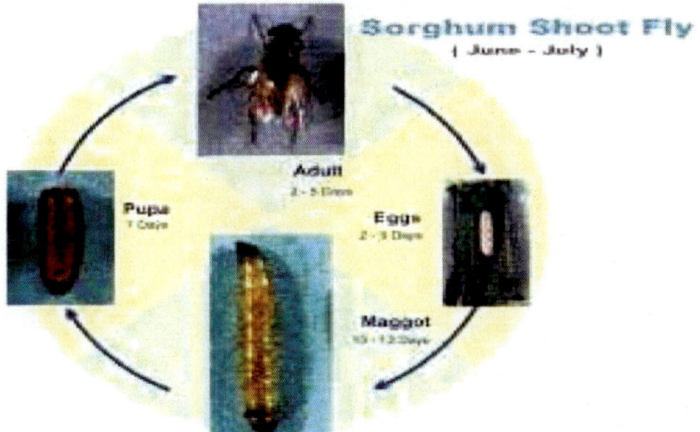

5. Maize Stem borer, *Chilo partellus* (Lepidoptera: Pyralidae)

This is also a major pest of sorghum.

6. Other minor pests

- Blister beetle (*Zonabris phalarata*) Coleoptera; Meloidae
- Mealy bugs (*Heterococcus rehi*) Hemiptera; Pseudococcidae
- White grubs (*Holotrichia consanguinea*, *Holotrichia insularis*, *Anomala bengalensis*) Coleoptera; Scarabaeidae
- Thrips (*Anaphothrips sudanensis, Sorghothrips jonnaphilus*, *Florthrips traegardhi*) Thysanoptera; Thripidae
- Grasshoppers, aphids, hairy caterpillars, leaf hoppers and borers.

RAGI

1. Pink borer, *Sesamia inferens* (Lepidoptera: Noctuidae)

Distribution

This is a polyphagous pest distributed throughout India and Pakistan. It is a common pest of Ragi, but minor pest of maize, sorghum, wheat, sugarcane and rice.

Damaging symptoms

Young caterpillars immediately after hatching, bore into the epidermal layers of leaf sheath. Later they bore into the stem, resulting in drying up of the growing shoot, thus showing typical dead-heart symptoms. As a result of their attack, young plants are killed, while the older plants produce a few grains only. When the attacked plants die, the larvae move to the adjoining plants.

Life cycle

These nocturnal moths lay eggs either on leaves or on ground. Eggs hatch in 6-8 days and the larvae mature in 3-4 weeks. Mature larvae have pinkish brown, smooth cylindrical body, measuring about 25 mm. They pupate inside the stem or in between the stem and leaves. Pupal stage is of about a week and the whole life cycle is completed in 6-7 weeks. Moths are stoutly built and straw coloured. 4-5 generations are completed in a year.

Management strategy

- Apply Phorate 10 G or Carbofuran 3 G in the central whorl of ragi as mentioned for maize stem borer on young ragi plants.

Pink borer larva Pupa Adult

2. Aphids: *Schizaphis graminum, Hysteroneura setariae, Rhopalosiphum maidis* (Hemiptera: Aphididae)

Damaging symptoms

Both nymph and adults suck the cell sap, resulting in yellowing of leaves. Presence of colonies of aphids on leaves and ears, usually near the central leaf whorls, can be easily noticed due to the presence of ants, which are attracted as a result of honey dew secretion.

Life cycle

S. graminum are greenish yellow aphids; *H. setariae* are brown coloured; *R. maidis* are yellow coloured with dark green legs. They reproduce parthenogenetically. Nymphs and females look alike and are wingless, except that the latter are larger in size. Males are winged. During the active breeding season, there are no males and the rate of reproduction is very high.

Management strategy

- Spray Dimethoate 30 EC or Carbaryl 50 WP, especially directing the nozzles towards the central leaf whorls or where the colonies are found.

Schizaphis graminum

Hysteroneura setariae

Rhopalosiphum maidis

3. Root aphid, *Tetraneura nigriabdominalis* (Hemiptera: Aphididae)

Distribution

The pest is common in South India.

Damaging symptoms

It infests root system from the vegetative phage till the flowering stage of the crop in irrigated ragi during May – July. Both nymph and adults suck the cell sap from the base of the plant, resulting in wilting and drying of plants in patches. Presence of ants at the base of the plant is the clear indication of its presence.

Life cycle

Adults are soft-bodied, pale green coloured or pinkish, globular aphids. They reproduce parthenogenetically. Nymphs and females look alike and are wingless, except that the latter are larger in size. Males are winged.

Management strategy

- Pull out plants which are stunted severely, with roots intact and destroy the aphids.
- Spray Dimethoate 30 EC or Carbaryl 50 WP, drenching the rhizosphere of the infested and adjoining plants.

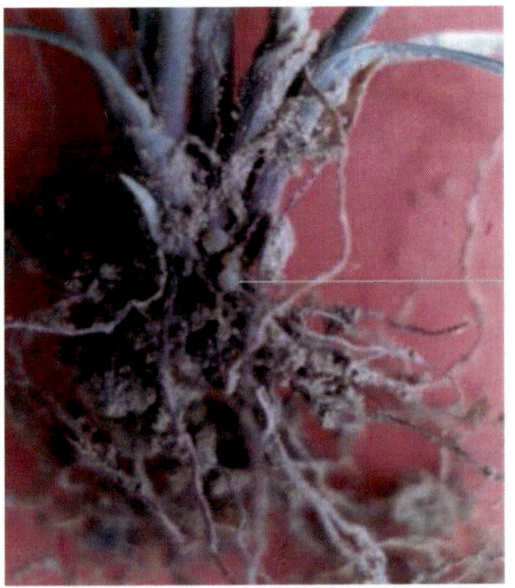

Ragi root aphid

4. Other minor pests

4. Ear head bug *Calocoris angustatus* (Hemiptera: Miridae)
5. Ash weevils *Myllocerus* spp.(Coleoptera : Curculionidae)
6. Cut worm *Spodoptera exigua* (Lepidoptera: Noctuidae)
7. Black Hairy caterpillar *Estigmene lactinea* (Lepidoptera : Arctiidae)
8. White grub *Holotrichia consanguinea* (Coleoptera: Melolonthidae)
9. White borer *Saluria inflicta* (Lepidoptera: Phycitidae).

PULSES

Black gram / Green gram

1. Pod borer, *Maruca testulalis* (Lepidoptera: Crambidae)

Damaging Symptoms

The larvae webs together the flowers and feeds on them and also bores into pods and feeds on the seeds causing extensive losses. It has also been noticed to web the leaves of dhaincha.

Life Cycle

Larva is greenish with brown head and short dark hairs on black warts on the body. It pupates in a thick silken cocoon on the pod or leaf fold. The pupa is yellow with a greenish body. Moth has dark brown forewings with a white cross band and white hind wings on a dark border.

Management Strategies

- Deep ploughing helps in burying the larvae/pupae too deep in soil to emerge or conversely expose them to predators, parasites or adverse weather conditions.
- Foliar application of dimethoate, metasystox, Chlorantraniliprole 18.5%SC, thiodicarb 75 WP, etc. may be given as per need.

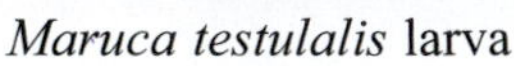

Maruca testulalis larva

M. testulalis adult

2. Hairy caterpillar, *Amsacta moorei* (Butler) (Lepidoptera: Arctiidae)

Amsacta albistriga (Walker)

Spilosoma obliqua Walker (Lepidoptera: Arctiidae)

Damaging Symptoms

Larvae defoliates, thereby affecting grain yield. Young larvae feed on growing points of the plant and on the photosynthetic part from the underside of leaves, while the later instars are voracious feeders and completely skeletonize the crop, in case of heavy infestation. This is a polyphagous insect and feeds practically on all kinds of vegetation during kharif.

Life Cycle

Full grown caterpillars are reddish amber to olive green with their body covered with numerous long hairs, arising from the fleshy tubercles. *A. moorei* moth is stoutly built having white wings with black spots. Outer margin of forewings, anterior margin of thorax and entire abdomen are scarlet red. While the head, thorax and under side of the body are dull yellow in *S. obliqua* moths.

The pest is active during rainy season (Mid –June till end of August) and passes the rest of the year in pupal stage in soil. Moths appear with first shower, and lay yellow spherical eggs in clusters of 700 – 850 each on the under surface of leaves. The young caterpillars after hatching in 2-3 days, feeds on the photosynthetic part of the leaves, thus they become whitish in colour. Such leaves are easily recognizable from a distance and can be plucked along with the congregated larvae and destroyed. Full grown larvae enter the soil and pupate at a depth of about 23 cm. They emerge next year at the onset of monsoon.

Management strategies

Cultural practices

- Timely sowing should be done.
- Field sanitation, rogueing.
- Crop rotation.
- Deep summer ploughing and Pre-monsoon deep ploughing (two/three times) to expose the hibernating pupae to sunlight and predatory birds.
- Destruction of alternate host plants.
- Eradicate weeds and volunteer sesame plants.

Mechanical practices

- Collection and destruction of insect infested plant parts.
- Collection and destruction of eggs and early stage larvae. The leaves with young congregated larvae can be easily collected and destroyed.
- Handpick the older larvae during early stages of the crop.
- The infested shoots and pods may be collected and destroyed.

- Handpick the gregarious caterpillars and the cocoons which are found on stem and destroy them in kerosene mixed water.
- Moths are strongly attracted to light. Setting up of light traps following the first shower of monsoon and continued throughout the period of emergence for about one month, greatly reduces pest incidence. Use light trap @ 1/acre and operate between 6 pm and 10 pm.
- Install pheromone traps @ 4-5/acre for monitoring adult moths activity (replace the lures with fresh lures after every 2-3 weeks).
- Erect bird perches @ 20/acre for encouraging predatory birds such as King crow, common mynah etc.

Use of Botanicals

- Conservation of natural enemies through ecological engineering.
- Spray neem oil @ 5 ml/l as foliar sprays.
- Application of 1.5 kg Carbaryl 50 WP or Quinalphos 25 EC @ 1.25 in 600-1000 L water /ha.

Hairy caterpillar damaged leaves

Hairy caterpillar egg clusters

Young caterpillars after hatching

Fully grown caterpillar

Adult moth

3. Stem fly *Ophiomyia phaseoli* (Tryon) (Diptera: Agromyzidae)

Damaging Symptoms

The larvae are leaf and stem miners. The maggots bore into the stem thus causing withering and drying of the affected shoots, thus reducing the yield. Adults also puncture the leaves, which thereby turn yellow. Damage is more severe in seedlings than in grown up plants.

Life Cycle

The adult metallic black flies are more active in summer. The female lays 14-64 elongate, oval, white coloured eggs, into the leaf tissue. The maggots after hatching in 2-4 days, start feeding on the leaf tissue initially, but later move to the terminal stem. The maggot undergoes pupation in 6-7 days in March-April and 9-12 days in November – December in the galleries made by them. This pest completes 8-9 generations in a year. It passes winter as larva or as pupa.

Management Strategies

- Timely sowing escapes the peak pest incidence.
- Removal and destruction of the affected branches during initial stages of attack.
- Seed treatment with phorate @ 1 kg per 50 kg seed gives protection for 2-3 weeks after germination.
- Foliar application of dimethoate, metasystox, monocrotophos, etc. may be given as per need.

Ophiomyia phaseoli adult fly

O. phaseoli various stages of life cycle

4. Whitefly, *Bemisia tabaci* (Genn.) (Diptera: Aleyrodidae)

Damaging Symptoms

Both adults and nymphs suck the cell sap, thus devitalizing the plant and are also vector of Mungbean Yellow Mosaic Virus (MYMV) and Urdbean Yellow Mosaic Virus (UYMV). They also excrete honeydew on which the sooty mould grows, interfering with the normal photosynthesis process. Consequently the crop gives sickly, black appearance from a distance. It transmits a number of viral diseases in several crops.

Life Cycle

Females lay approximately 119 eggs singly on the underside of the leaves. The elliptical, stalked, yellowish coloured eggs hatch in 3-5 days in April – September, 5-17 days in October–November and 33 days December–January. The nymphs suck cell sap and moults thrice to form pupae within 9-14 days in April–September and 17-81 days during winter months. The total life cycle is completed in 14–122 days and there are 11 generations in a year.

The nymphs are louse-like sluggish creature, while the adult are pale yellow, 1.0-1.5 mm long, with white waxy powdery covering.

Management Strategies

- Early sowing during spring (15th March) suffers less damage.
- Inter/mixed cropping with taller crops like maize, sorghum, cotton, pearl millet, pigeonpea, etc., shields the crop, thereby checking the dispersal of flying insects.
- Spraying the crop at bud initiation stage with Dimethoate, Malathion, Metasystox or Phosphamidon effectively controls this pest.

Adult *Bemisia tabaci*

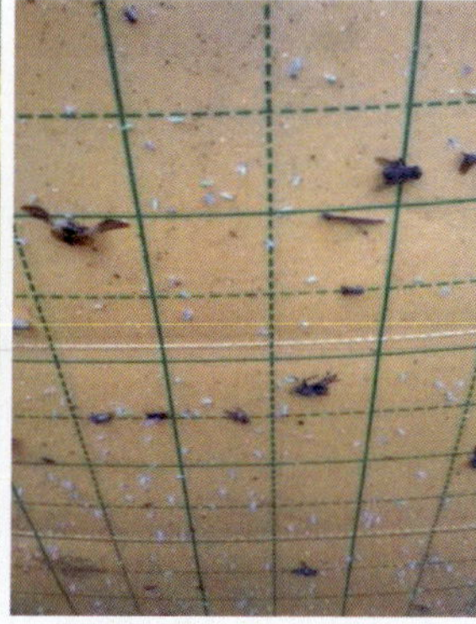

B. tabaci sticking to yellow sticky trap

Yellow mosaic affected urdbean

5. Thrips, *Caliothrips indicus* (Bagn.) (Thysanoptera: Thripidae)

Damaging Symptoms

Both nymphs and adults damage the flowers and due to their damage the flowers shed before opening. Infested inflorescence becomes abnormal showing symptoms of flowers drop. In case of severe incidence there is 100% loss in grain yield.

Life Cycle

Adults are slender about 1 mm in length. The nymphs resemble adults, but are wingless and slightly smaller. Adult female lays kidney shaped eggs singly in the slits. Eggs hatch in 4-9 days and nymphs start feeding on plant juices by lacerating the leaf tissues. Nymphs pass through four stages and they pupate in 4-6 days in the ground at a depth of 25 mm. Several generations are completed in a year.

Management Strategies

- Early sowing during spring (15th March) suffers less damage.
- Timely irrigation in summer crop protects from thrips damage as their multiplication is very rapid under drier weather conditions.
- Chemical control same as for whiteflies.

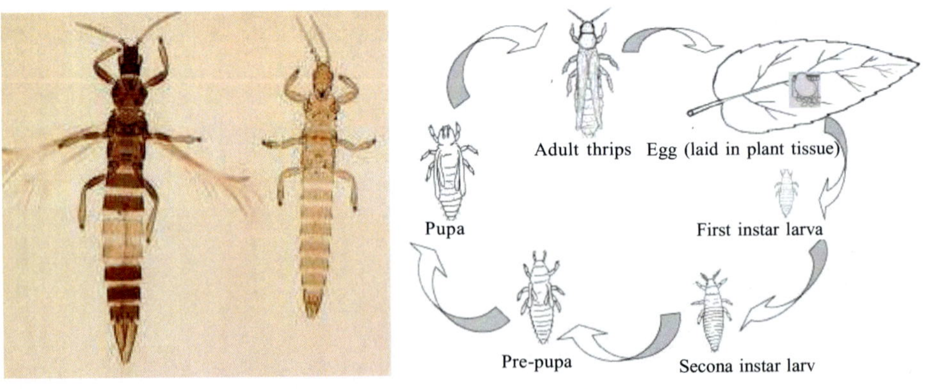

Caliothrips indicus adult

C. indicus life cycle

6. Green Jassid, *Empoasca kerri* (Hemiptera: Cicadellidae)

Damaging Symptoms

Both adults and nymphs suck cell sap. The attacked leaves turn rust-red, show mottling, whitening and yellowing, cup downwards, dry up and fall to the ground.

Life Cycle

These are elongate (3 mm long), active, wedge shaped, green insects generally found on the under surface of leaves. Front wings are long, narrow, semi-transparent and pale green in colour with grey distal regions. A single female lays 25-60 eggs in 25-30 days. Eggs hatch in 4-11 days. Nymphal period is of 25 days, while the adult male and female longevity is 2 weeks and 13 weeks respectively. It breeds throughout the year.

Management Strategies

- Inter/mixed cropping with taller crops like maize, sorghum, cotton, pearl millet, pigeonpea, etc., shields the crop, thereby checking their dispersal.
- Chemical control same as for thrips.

Empoasca kerri eggs, nymphs and adult

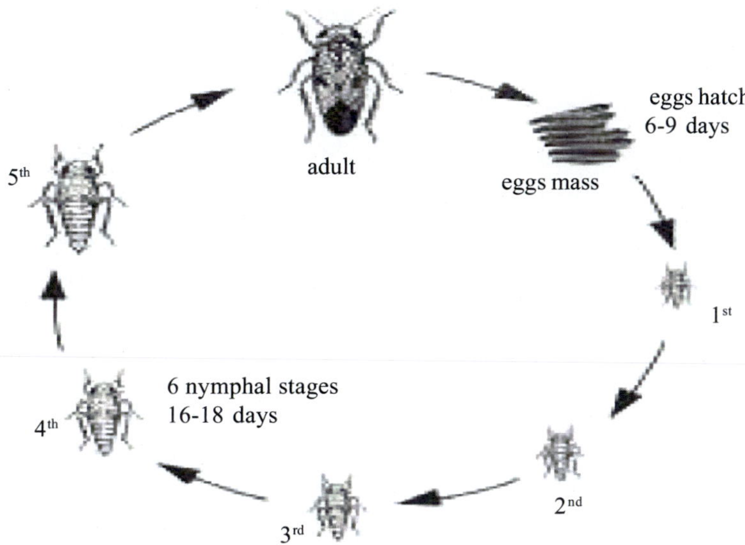

Empoasca kerri life cycle

7. Bean bugs, *Clavigralla gibbosa* Spinola (Hemiptera: Coreidae), Dusky cotton bug, *Oxycarenus laetus* Kirby (Hemiptera: Lygaeidae)

Damaging Symptoms

Both nymphs and adults suck cell sap from leaves, stem, pods, flower buds. Pods show yellow patches and later pods as well as grains shrivel in case of heavy attack. The grain size reduces thereby reducing the yield significantly.

Life Cycle

C. gibbosa

Adult bugs are greenish brown in colour, 20 mm long with spined pronotum and swollen femur. A single female lays as many as 52 eggs usually on pods and sometimes on leaves and other floral buds. The mean egg, nymphal and adult periods of *C. gibbosa* on field bean is 8, 20 and 11 days. The total life cycle of *C. gibbosa* is 42 days. The pest is active from middle of October till end of May and has six overlapping generations during this period.

O. laetus

The female bug lays transparent, light yellow, oval or cigar shaped eggs either singly or in groups of 2 - 10 usually on pods. The freshly hatched light pink coloured nymphs hatch in 8 days with a small black tail like process. The total nymphal period ranges from 15 to 23 days. Freshly emerged adult is bright red in colour. Body colour of adult turns to dark brown after four to five hours of moulting. The adult bugs are small, flat with pointed head and uniform dark brown body with dirty white or dusky brown transparent hemielytra. Total life cycle of varies from 35 to 50 days.

Management Strategies

- Shaking the infested plants over vessels of oil and water or oily cloth gives very effective control.
- Egg masses of bugs should be collected and kept away from the cropped area as 30% of the egg masses are parasitized by *Gryon* sp. This practice will allow the parasite to be released in nature but will prevent the nymphs from causing damage.

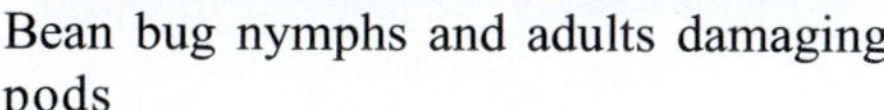

Bean bug nymphs and adults damaging pods

Bean bug adult

8. Bean mite, *Polyphagotarsonemus latus* (Banks) (Acarina: Tarsonemidae)

Damaging Symptoms

Both nymphs and adults suck cell sap from leaves, stem, pods, flower buds. Reduction in photosynthesis and instability of water balance are some of the damaging effects to plants. As a result terminal leaves and flower buds become cupped and distorted. On the underside on the leaves corky brown areas appear near the mid vein. Pods as well as grains shrivel in case of heavy attack.

Life Cycle

Eggs are generally laid on underside of the leaves, which hatch in 2-3 days. Nymphs are very small, pear-shaped and have three pairs of legs. The larvae feed for 1 to 3 days before going into the resting pupal stage, which lasts for 2-3 days. The life cycle, from egg to adult, is completed in 4 to 6 days. Adult mites are very small, elliptical in shape, light, translucent yellowish green in colour. Females live for about 10 days and lay an average of 2 to 5 eggs per day (20-50 eggs per female).

Management Strategies

- Chemical control same as for thrips.
- Chemicals should be carefully applied and care should be taken to not interfere with natural enemies that offer biological control of other members of the pest complex.

Polyphagotarsonemus latus adult

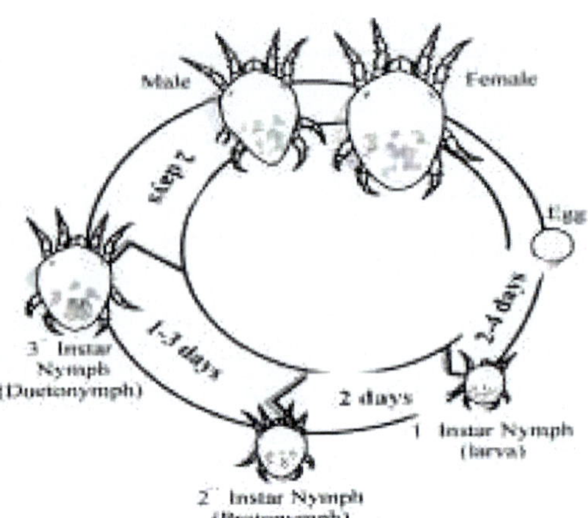

P. latus life cycle

9. Cut worm, *Agrotis ipsilon* (Hufnagel), *Agrotis flammatra* Schiff. (Lepidoptera: Noctuidae)

Damaging Symptoms

Larva become active at night and damages the seedlings by cutting at the base or just above soil surface. The cut plants or twigs are dragged and partially buried into the soil. This indicates their presence in the field. During day, the caterpillars remain hidden in cracks or within clods.

Life Cycle

A single female moth can lay as many as 639-2252 eggs on earthen clods, chickpea stem bases or on both sides of leaves. The average incubation, larval, pupal and total duration varies from 2.7–5.1, 18.2–39.5, 10.3–25.3 and 31.4–69.8 days respectively. Extremely dry weather during April–May adversely affects cutworm multiplication. During off-season (summer and rainy), the pest remains scattered among weeds growing nearby.

Management Strategies

- Field ploughing 4-5 times before sowing kharif crops breaks the clods in the field and reduces pest incidence.
- Clean cultivation, i.e. cleaning the weeds from bunds, nearby fields.
- Braconids, *Microgaster* sp., *Bracon kitcheneri* and *Fileanta ruficanda* parasitizes *Agrotis* larvae, while *Broscus punctatus* and *Liogrullus bimaculatus* predate on cutworms.
- This pest can be controlled by applying Lindane 2 per cent dust @ 125 kg/ha before sowing of Rabi crop.

Cut worm larva

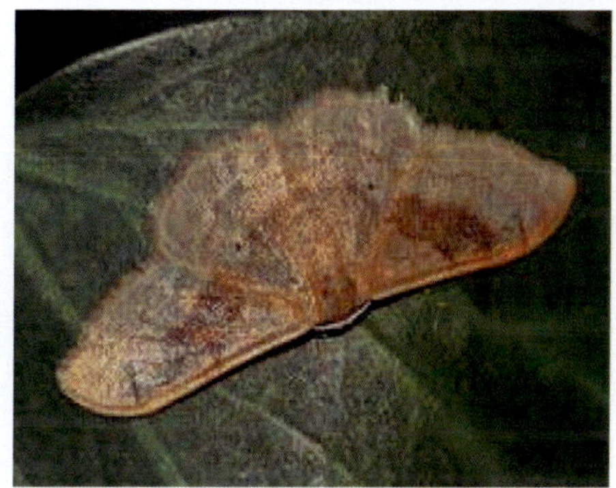

Cut worm adult

Life cycle

FRENCH BEAN

1. Hairy caterpillar, *Amsacta moorei* (Butler) (Lepidoptera: Arctiidae)

Damaging symptoms, life cycle and management strategies

- Same as for insect pests of urdbean.

2. Blister beetle, *Mylabris pustulata* (Thunberg) (Coleoptera: Meloidae)

Damaging Symptoms, life cycle and management strategies

- Same as in case of maize.

3. Bean fly *Ophiomyia phaseoli* (Tryon) (Diptera: Agromyzidae)

Damaging Symptoms

Maggots mine sub-epidermally through the leaves. Later they tunnel the stem, resulting in drooping of the first two leaves and yellowing. Plants are most seriously affected at the seedling stage. Older plants also suffer leaf fall and stunting but are not usually killed.

Life Cycle

It is a very minute insect, shiny black in colour. Eggs, wing veins are light brown in colour. Females lay slender white eggs singly in holes made on upper leaf surface of young leaves. On hatching the larvae mine underneath the epidermis, until it reaches the midrib and then to the leaf stalk and stem. Black coloured barrel shaped pupae, pupate inside the stem. Total life cycle is completed in 2-3 weeks. As many as seven generations are completed during the season.

Ophiomyia phaseoli adult

Management Strategies

- Timely sowing minimizes the attack.
- Seed treatment with phorate @ 1 kg per 50 kg seed gives protection for 2-3 weeks after germination.
- Soil application of 10 kg Phorate 10G is effective upto 40 days of sowing.
- Foliar application of dimethoate, metasystox, monocrotophos, etc. may be given as per need.

4. Bean thrips, *Megalurothrips distalis* (Karny) (Thysanoptera: Thripidae)

Damaging Symptoms

Both nymphs and adults damages the flowers and due to their damage, flowers shed before opening. Infested inflorescence becomes abnormal showing symptoms of flowers drop. In case of severe incidence there is 100% loss in grain yield.

Life Cycle

Females are active flier and seek tender leaves on the host-plant for oviposition. The egg is oval, milky white, opaque or translucent Oviposition period lasts from 3-4 days after adult emergence and continues for 10-15 days. Incubation period of eggs varies from 2 to 4 days. The newly hatched, pale, yellowish white first instar larva feeds mainly in leaf blade tissue before entering midrib eventually entering the stem. Before pupation, the fully grown larva ceases feeding for 6-10 hours, constructs a semicircular window in epidermis for escape of adult emerging from the pupae. It pupates beneath the epidermis of the stem, near the soil surface. When newly emerged, it is yellow with a brownish tinge and distinctly darker ends.

Management Strategies

- Spraying the crop at bud initiation stage with Dimethoate, Malathion, Metasystox or Phosphamidon effectively controls this pest.

Megalurothrips distalis adult

M. distalis life cycle

5. Cut worm, *Agrotis ipsilon* (Hufnagel), *Agrotis flammatra* Schiff. (Lepidoptera: Noctuidae)

Damaging symptoms, life cycle and management strategies

- Same as for insect pests of urdbean

6. Bean mite, *Polyphagotarsonemus latus* (Banks) (Acarina: Tarsonemidae)

Damaging symptoms, life cycle and management strategies

- Same as for insect pests of urdbean

OILSEEDS

Til (Sesame)

1. Hairy Caterpillar, *Amsacta moorei* (Butler)
Amsacta albistriga (Walker)
Spilosoma obliqua Walker (Lepidoptera: Arctiidae)

Damaging symptoms, life cycle and management strategies

- Same as for insect pests of urdbean.

2. Leaf hopper: *Orosius albicinctus* Distant (Hemiptera: Cicadellidae)

Damaging symptoms

Both nymphs and adults suck the sap from leaves and transmit phyllody disease. Curling of leaf edges and leaves turn red or brown. The leaves dry up and shed. In the phyllody disease affected plants, there is no pod formation at all.

Life cycle

It is the vector of a virus that causes a phyllody disease of sesamum and also infests other important crop plants in India. In the field, numbers are low in winter and high in summer; in January-March, only adults are found. Egg stage lasts for 6 days in June and 96 days in December. There are five nymphal instars and the duration of nymphal development 12 days.

Management Strategies

- Use yellow sticky traps @ 50 trap/ha.
- Follow common cultural, mechanical practices and botanical spray.

Orosius albicinctus adult

Phyllody disease

- Sow / plant sorghum/maize/bajra in 4 rows all around sesame crop as a guard/barrier crop.
- Oxydemeton–methyl 25% EC@ 480 ml in 200-400 l of water/acre.

3. Aphids

Damaging Symptoms

Both the nymphs and the adults pierce the plant tissues to feed on plant sap. Infected leaves become severely distorted when the saliva of aphids are injected into them. Heavily infested ones turn yellow and eventually wilt because of excessive sap removal. The aphids' feeding on the plant causes crinkling and cupping of leaves, defoliation, and stunted growth. Aphids produce large amounts of a sugary liquid waste called honeydew, on which a fungus develops called sooty mold, turning leaves and branches black. The appearance of a sooty mold on plants is an indication of an aphid infestation.

Life Cycle

The eggs are very tiny, shiny black, and are laid in the crevices of bud, stems, and barks of the plant. Nymphs look like young adults, mature within 7-10 days, and are then ready to reproduce. Adults are small, 3-4 mm long, soft-bodied insects with two projections on the rear end and two long antennae. Their body color varies from yellow, green, brown, to purple. Females can give birth to live nymphs as well as can lay eggs. However, the primary means of reproduction for most aphid species is asexual, with eggs hatching inside their bodies, and then giving birth to living young. Winged adults, black in color, are produced only under adverse conditions, when it is necessary for the colony to migrate, or there is either overcrowding in colonies, or unfavorable climatic conditions.

Management Strategies

- Control and kill ants. Cultivate and flood the field. This will destroy ant colonies and expose eggs and larvae to predators and sunlight. Ants use the aphids to gain access to nutrients from the plants.
- Avoid using heavy doses of nitrogen fertilizers, as this result in more vigorous vegetative growth, and aphids love succulent tender leaves. Split application of nitrogenous fertilizers should be done during seedling and flowering stage.
- Installation of Yellow sticky traps @ 50 / ha, attracts them and they get stick to it. These traps can be prepared by using petroleum jelly or used motor oil on yellow plywood, 6 cm x 15 cm in size or cardboard with

yellow glossy sheets pasted on both sides. Traps when hung should be positioned 61 cm zone above the plants.

- Follow common cultural, mechanical practices and botanical spray.
- Other practices same as for sesame leaf hopper.

Aphid nymph

Coccinellid adult feeding on aphids

4. Whiteflies, *Bemisia tabaci* (Hemiptera: Aleyrodidae)

Damaging Symptoms

Both larvae and adults pierce and suck the cell sap from leaves. Their feeding causes yellowing, drying, premature dropping of leaves, weakening and early wilting of the plant, ultimately resulting in plant death. Whiteflies also produces honeydews like aphids that serve as the substrates for the growth of black sooty molds on leaves and fruit.

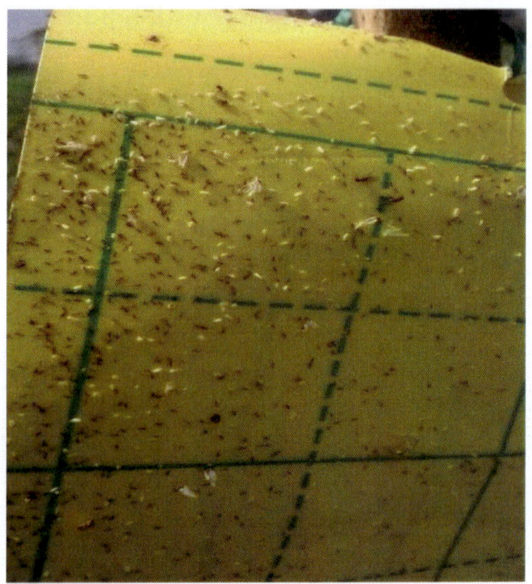

Whitefly adults sticking to Yellow sticky

Life Cycle

The eggs are tiny, oval-shaped, white in colour, about 0.25 mm in diameter, and stand vertically on the leaf surface. They are deposited on the underside of leaves, sometimes in a circle or oval-shaped patterns. The larvae are transparent, ovate, and about 0.3-0.7 mm in size. The pupae are dirty-white and surrounded by wax and honeydews. The adults are about 1mm long with two pairs of white wings and light yellow bodies. Their bodies are covered with waxy powdery materials. A female can produce as many as 200 eggs in her lifetime and mating is not necessary. It takes about 40 days to develop from egg to adult.

Management Strategies

- Same as for sesame aphids.

5. Leaf webber or roller and capsule borer, *Antigastra catalaunalis* Duponchel (Lepidoptera: Crambidae)

Damage symptoms

Young larvae feed externally by making a loose web, which folds several leaves together. The larvae feed on leaves and young shoots. That web gets filled with the excreta (frass). At later stage, the larvae infest the sesame fruit capsule making an entrance hole on the lateral side and feeding on the seeds inside the capsule leaving excreta on the seeds.

Life cycle

Minute conical eggs are laid singly on the under surface of leaves, on capsules and branches. Eggs are white in colour, which later change to dark white before hatching. Larva is a cylindrical caterpillar, which passes through five larval instars before going in to pupation. The larva is greenish in colour with black head and about 15 mm in length. Pupa is slender, long necked and greenish-reddish-brown in colour. The body is light reddish brown, greenish white at first. It changes gradually to pale reddish white, dark reddish, reddish brown and pale whitish later on. Adult is medium sized moth with reddish yellow forewings.

Management Strategies

- Follow same cultural, mechanical practices and botanical spray.
- Quinalphos 25% EC @ 800 ml in 200-400 l of water/acre.
- Carbaryl 10% D.P. @ 10 Kg/acre.

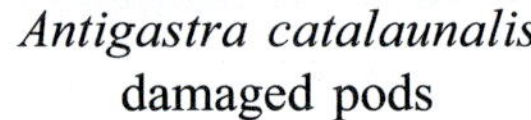

Antigastra catalaunalis damaged pods

Adults

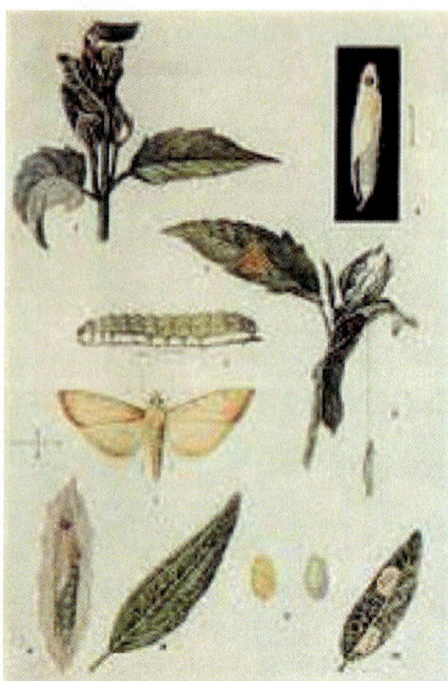

Various stages of *A. catalaunalis*

6. Gall fly: *Asphondylia sesame* Felt (Diptera: Cecidomyiidae)

Damage symptoms

Maggots feed inside the floral bud, leading to formation of gall like structure which do not develop in to flower/capsules. The affected buds wither and drop.

Life cycle

The maggot is white to orange in colour, legless and with body tapering exteriorly. It grows up to 3 to 4 mm in length. Maggots feed inside the floral buds and young capsules leading to formation of galls of up to 6 mm in diameter. Maggot pupates inside the galls. The adult is a 5 mm long red-bodied midge (mosquito-like fly). Female midges lay eggs along the veins of terminal leaves.

Asphondylia sesame damaged pods

Management strategies

- Same as for sesame leaf roller.

  Natural enemies of gall fly:

  Parasitoids: *Pteromalus fasciatus* etc.

  Predators: Spider, ladybird beetle, lacewing etc.

7. Sphinx moth or Dead head moth: *Acherontia styx* (Lepidoptera: Sphingidae)

Damaging symptoms

The young larvae roll together a few top leaves and feed them. The damage is caused by the larvae which feed voraciously on leaves and defoliate the plants. At flowering, larvae feed inside the flowers and on capsule formation, larvae bore into capsule and feed on developing seeds. The moth is also harmful as it sucks honey from the honey combs in apiaries.

Life cycle

Eggs are globular is shape and are laid singly on the under surface of leaves. The egg period is 2-5 days. Larva is stout, green with yellowish oblique stripes and curved anal horn. The larval period lasts for 60 days. It pupates in earthen cocoon in soil. The pupal period lasts for 14-21 days and 7 months in summer and winter, respectively. Adult moth is giant hawk moth, brownish with a characteristic skull marking on the thorax and violet yellow bands on the abdomen. Hind wings are yellow with black markings.

Management Strategies

- Deep ploughing exposes the pupae for predation to insectivorous birds.
- Hand picking (collection) and destruction of caterpillars.

Natural enemies of sphinx moth

Predators: Lacewing, ladybird beetle, reduviid bug, spider, red ant, black drongo (King crow), common mynah, big-eyed bug (*Geocoris* sp), pentatomid bug (*Eocanthecona furcellata*), preying mantis etc.

Acherontia styx caterpillar

A. styx adult

POLYPHAGOUS PESTS

1. White grub *Holotrichia consanguinea* Blanchard (Coleoptera: Scarabaeidae)

Host: This pest is polyphagous pest and feed on almost all the *kharif* crops like maize, jowar, shorghum etc.

Distribution: This is worldwide importance as noxious crop pest and cosmopolitan in occurrence. White grub has attained a status of major pest of a variety of rainy season crops in India. i.e. Rajasthan, Gujarat, Maharashtra, Karnataka, T.N., Bihar, U.P., Jharkhand and Odisha.

Nature of damage: The damage is caused by grubs and adult beetle feeds on fine root lets, nodules and main root. Adults feed on shrubs and trees growing nearby the cultivated fields by defoliating them. Plants damaged by the grub, gives a wilted appearance and finally dries up. The damage caused by this pest estimated to the tune of 40-80%.

Life cycle: The life cycle of this pest passes through the four stages viz. egg, grub, pupa and adult.

Egg: The beetle emerge from the soil after dusk (7.30-8.00 pm) followed by good pre-monsoon rains, which may occur from April to June. After feeding and mating the beetle get back to the soil in early morning hours and lay eggs. This is generally laid singly in loose soil or in an earthen cell inside the soil upto the depth of 10cm. The eggs are oval, creamy white when freshly and later turn to brown in

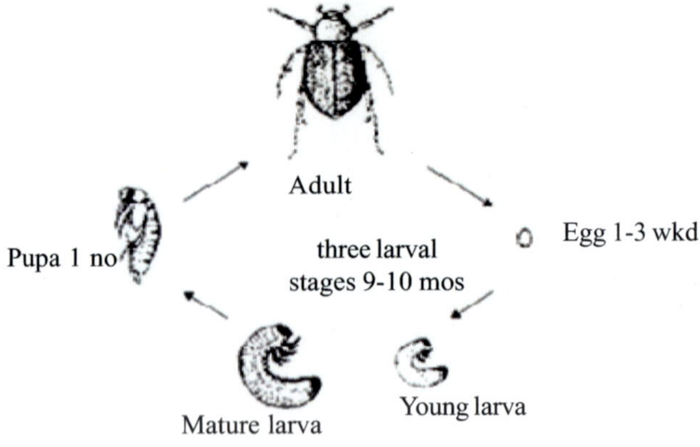

colour. About 30-120 eggs are laid by single female and hatching period about 7-14 days.

Grub: Newly hatch grub in creamy white, measure 10-12 mm in length and 2-2 mm width, and feed on organic matter. There are three larval instars and full grown grub is curved 'C' shaped and dirty white in colour. Head is dark brown with strong mandibles and prominent thoracic legs. The larval period for the first and second instar 9-12 days and 3-4weeks, respectively, while in 3rd instar about 6 weeks. Thus total larval period varies from 72-82 days.

Pupa: The full grown grubs move down deeper in the soil in search of moisture and for pupation. Its constructs an earthen cell in which it passes a quiescent or prepupal stage which lasts for 1-6 weeks. Pupa is light yellow and extremely tender, but as it grows older it turn brown in colour. This period varies from 10-27 days depending upon the type of soil.

Adult: Newly formed beetle is cream in colour with soft white elytra with the lapse of time the colour changes to brown and elytra gets hardened. Generally the adults are lamellate and males being smaller than females. Beetles remain within the earthen cell and do not come out of soil until the onset of summer rains. Total life cycle is completed from 90-108 days and all the known species in India have one generation in a year.

Management

- The use of light traps is the successful method for the collection of beetles during emergence in the night.
- The repeated ploughings, preferably soon after the summer rains, help in exposing the various stages of white grubs to their natural enemies.
- Biological control or natural control such as *Bacillus thuringiensiis, B. popillae* (bacteria) and *Beauveria bassiana, Aspergillus parasiticus* (fungus).
- Adult beetle controlled by the solid application i.e. Phorate 10g @ 25 kg/ha or Carbofuran 3 g @ 35 kg/ha.
- Spray of Chlorpyriphos 20EC @ 5 liters per ha, offers the economic and effective control of grubs.

2. Rodents

Rodents are one of the most important non - insects pests of agricultural crops, particularly *kharif* and *rabi* cereal.

Types of rodents

- Lesser bandicoot rat: *Bandicota bengalensis*
- Field mouse: *Mus booduga*
- Indian gerbil: *Tatera Indica*
- Soft furred field rat : *Rattus meltada.*

Damage at different stages: In India, rodents have been estimated to cause 5 to 10% losses in cereal crops. Among the field crops, rice and wheat are the most vulnerable crop to rodents. In addition to tiller cutting, they also hoard ripened panicles inside their burrows.

Nursery in rice: The nurseries are drained out and the rodents run freely inside the bed spoiling all germinated seed. Later, they also cut the seedlings 1-2 inches above the water level.

Main field: Sometimes the rodents pull out the transplanted seedlings in paddy and create gaps in the main field. Generally, their activity is confined to inside field leaving 2-4 meters on all sides of the field. In the initial stage, damage appears in patches and after some time, all these small patches become into one big patch. Damage increases with the onset of panicle initiation and continues up to panicle emergence.

Management of Rodents

- In Local traps called 'butta' are extensively used for the control of rodents in cereal. These traps provide fairly good results when applied after chemical control operation.
- However when directly used, trapping will be costly affair and one cannot manage entire population over large areas.
- Moreover, at certain crop stages, like primordial formation, rodents are not attracted towards traps.

Natural Smoke

- The main principle involved in this operation is simply filling the burrows with smoke, which causes suffocation to rodents ultimately leading to their death.
- The smoke liberated by burning rice straw mainly contains carbon-di-oxide.

Chemical Control Fumigation

- The fumigants like Aluminum phosphide is effective and widely used for the control of field rodents living in burrows.
- The control of rodents using rodenticides is the more common way.

1. Acute rodenticides (Single dose and quick acting), *Eg* : Zinc phosphide.
2. Chronic rodenticides (Multi dose and slow acting), *Eg* : Warfarin, Bromodiolone.

1) Acute Rodenticides

Among the acute rodenticides, Zinc phosphide and Barium carbonate are registered for use.

Action plan for Rodent control

Day 1 Identify live burrows and place 20 g of pre-bait Material inside the burrow.

Day 3 Place 10 g Zinc phosphide poison bait inside the Burrow.

Day 4 Collect dead rats and bury them. Close all the Burrows.

Day 5 Eliminate the residual population through trapping or Burrow fumigation with burrow fumigator. Treat the opened burrows with aluminum phosphide 2 pellets per burrow.

Day 13 In dry black soils fumigation will not give results. Hence apply Bromodiolone 1 cake per burrow. Repeat Bromodiolone baiting.

Advantages of Zinc Phosphide

1. Quick killing.
2. Small quantity of chemical is required.
3. Single feeding.
4. Population can be brought down immediately.

Disadvantages of Zinc Phosphide

1. Necessity of prebaiting.
2. Low killing around 40 - 50 %.
3. Induce bait shyness.
4. Toxic to non target species.
5. Chances of secondary poisoning are more.

2) Chronic Rodenticides

- In order to overcome limitations and hazardous nature of acute rodenticides, lengthy baiting programme and possibility of resistance, new series of rodenticides have been developed and known as single dose anti-coagulants or second generation anti-coagulants. These rodenticides (Bromodiolone) combines better qualities of acute and chronic rodenticides.

- For effective and successful rodent control, the following programme should be adopted on large areas at a time on community approach.
- Chronic or multi-dose rodenticides (at present only anti-coagulants) are much safer than acute rodenticides because they are less toxic to the non-target species. But they have to be fed for 5 - 7 days to obtain desired results, which increases the cost of operation.

Day-1: Identify live burrows and place 15 g bromodiolone concentrate bait inside the burrow.

Day-2: Repeat Bromodiolone baiting in active or live burrows.

Day-3: Eliminate residual population through trapping or fumigation with burrow fumigator.

Principles

- Grow same maturity group cultivars on large areas to restrict the availability of the vulnerable stage (Reproductive) of the crop.
- Reduce the number and size of the bunds, keep them clean to locate burrows and avoid harborage.
- Rodent control operations should be taken up on large area at a time.
- It checks cross infestation or migration of rodents from untreated fields to treated fields.
- All the control operations should be completed before the crop attains primordial initiation stage since at this stage the rodents are invariably attracted to the rice crop.
- Rodenticides should be made available before beginning of the season.

Indian field mouse, *Mus booduga*

Soft furred field rat, *Millardia meltada*

Station baiting

Burrow baiting

Section - B
Major Diseases of Rainfed Kharif Crops

2

Cereals

MAIZE

1. Turcicum leaf spot

Causal organisms: ***Exserohilum turcicum***

Nature and Recurrence of disease: The occurrence of disease in all the growing regions. Warm weather is favourable for disease when leaves are wet from dew or rainfall for extended periods. Under favourable weather conditions, it can cause significant yield losses in susceptible hybrids. The fungus survives on maize residues, and the spores are spread long distances by wind (air borne). The fungus survives on diseased residues, volunteer maize and grasses.

Symptoms

- Long, elliptical, grayish green or tan lesions (2.5-1.5cm) appear on lower leaves progressing upward.
- Lesions on the leaves elongated between the veins, tan with buff to brown borders.
- Lesions size may vary in inbreds and hybrids due to different genetic background.

Management

- Spray the diseased crop with Mancozeb @2.5g/lt of water. Spray Zineb @ 2.5-4.0 g/lt of water (2- 4 applications) at 8-10 days interval.
- The crop debris should be ploughed down.
- Also, the resistant cultivars should be grown.

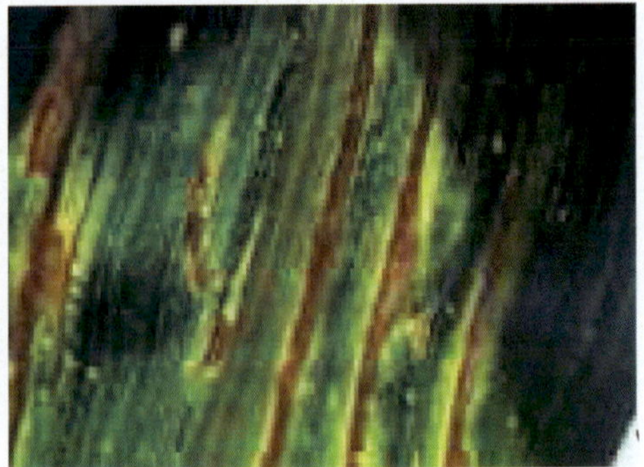

Turcicum leaf blight

2. Common rust

Causal organism: ***Puccinia sorghi***

Nature and Recurrence of disease: The inoculum survives in the form of uredospores on maize plants, which are generally grown round the year in different regions. By means of uredospores the disease cycle in repeated several times in the same season. The rust does much harm, especially in thickly planted maize during periods of high humidity.

Symptoms

- It appears at the time of tasseling.
- The circular to elongate, golden brown to cinnamon brown pustules are visible over both leaf surfaces changing to brownish black at plant maturity.
- The leaves become covered with numerous uredosori, loose their green colour, drop and dry up, and in severe causes the plant does not form the cobs.
- The uredosori remain scattered or in groups on both surfaces of the leaves, at first as little yellow points, which elongate and sometimes coalesce, turn reddish-brown and become pustular.
- The pustules burst soon, dislodging the reddish-brown powdery spore-mass.
- The teleutosori appear later, as larger black patches, upto 2mm long. They remain long covered by the host epidermis but burst eventually.

Management

- Spray of mancozeb@ 2.5g/lt of water at first appearance of pustules.
- Prefer early maturing varieties.

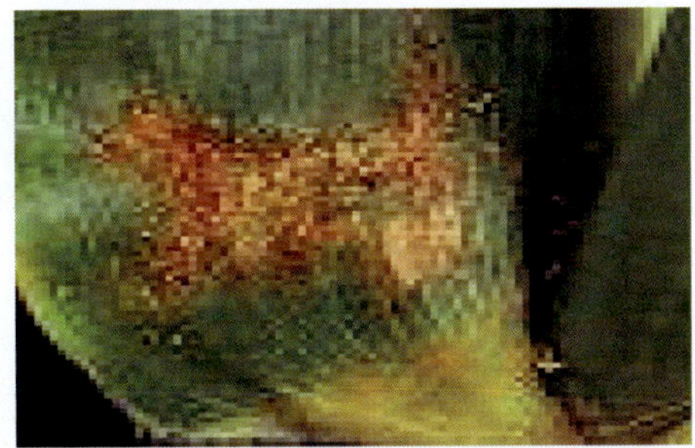

Common rust in maize

3. Leaf spot or Leaf blight

Causal organism- *Helminthosporium turcicum* Pass. *(Drechslera turcica* (Pass) Subram and Jain, *Trichometasphaeria turcica)*

Recurrence

The disease survives on plant debris in the soil-soil borne.

Distribution

The disease is quite common in many parts of India. It was first described in Italy in 1876, and since then has been recorded from the United States, South Africa, Japan and Philippines and most maize-growing countries.

Symptom

- In the beginning the typical symptoms is appearance of small water soaked lesions or spots on the leaf blade.
- These spots are linear to irregular in shape.
- These may extend into leaf sheath. The colour of the lesions are light olivaceous to brown which finally turns black to straw coloured as the tissue dry up.
- Diseased plants look burnt or frost bitten.

Management

- Clean cultivation always helps to reduce the disease.
- Spray the crop with Dithane M-45 or Zineb @ 0.2% at the appearance of disease. Spraying should be repeated after 15 days according to the severity of disease.

Leaf spot/ Leaf blight

4. Head smut

Causal organism: *Sphaecelotheca reiliana* (Kuhn) Clint

Recurrence of disease: The spores are shed from diseased plants on the soil around them and remain alive till the following season. The main infection comes from the soil and takes place usually during the early stage of seedling growth.

Distribution: In India, this disease occurs chiefly in Kashmir, the Punjab and Maharashtra. It has been known from India under the name *Ustilago pulveracea* Cke. It is known in the United States, Southern Europe, Africa, Australia and Japan.

Symptom

- The disease appears with the formation of tassels and ears where partial or complete transformation of the inflorescence into sori may be observed.
- The tassel is partially or wholly converted into smut spores. Finally whole cob is replaced with black spore mass and there is complete grain loss.

Management

- Sanitation.
- Crop rotation for 2-3 years.
- Uproot the affected plants and destroy.
- Seed treated with Cerasan or Emisan at the rate of 2g per Kg of seed.

Head smut of maize

5. Common smut or Maize smut

Causal organism: *Ustilago maydish*

Distribution

In India, this disease occurs chiefly in Kashmir, the Punjab and Maharashtra. It has been known from India under the name *Ustilago pulveraceae* Cke. It is known in the United States, South Europe, Africa, Australia and Japan.

Recurrence of disease: The spores are shed from diseased plants on the soil around them and remain alive till the following season. The main infection comes from the soil and takes place usually during the early stage comes from the soil and takes place usually during the early stage of seedling growth.

Symptom

- Characteristic galls are formed on the infected tissues.
- The galls may appear on the stem leaves axillary buds parts of the male and female flowers and more rarely on leaves. The galls are light coloured in the beginning and later on becomes dark and on rupturing expose the black powdery mass of spores.

Management

- Treat the seed with Captan @ 3g per kg of seed or carbendazim (50wp) at the rate of 2g per kg of seed.
- Use resistant varieties.

Common smut or maize smut

6. Bacterial stalk rot

Causal organism: *Erwinia corotovora f. zeae* Sabet

Distribution

Wide host range, including monocots as well as dicots. Infection sites are stomata, hydathodes, or the wounds of leaves or stalks. Overwinters in stalk residue on the soil surface line and transmitted by larvae of the maize borer occurs midseason after pollination of corn. Top rot can occur in plants that are sprinkler irrigated during periods of rapid growth.

Symptom

- The plants may be infected at any stage of growth in rainy season.
- The symptom appears at lower nodes and remain confined at one or two internodes only.
- The inner tissues also get infected in severe cases. The infected tissues are at first soft but later on they turn into dry mass of shredded fibers.
- The plants may topple down at this stage. The fields affected with the disease emit a typical vinegar (sirka) smell.

Management

- Rogue out the affected plants and destroy it.
- Drenching near the collar region with Mancozeb @ 2.5kg+50g streptocycline in 1000 L of water.

Bacterial stalk rot

7. Banded leaf and sheath blight

Symptoms

- Development of White lesions on leaves and sheath.
- Purplish or brown horizontal bands present on white lesions characterize the disease.

Management

- Seed treatment with peat based formulation (*Pseudomonas fluorescence)* @ 4.6 g/kg of seed treatment or as soil application @ 7g/lt of water (soil drenching) or foliar spray of Sheethmar (Validamycin) @ 2.7 ml/lt water provides effective control of the disease.
- Stripping of 2 lower leaves along with leaf sheath also gives effective control of the disease.

Banded leaf and sheath blight

8. Fusarium stalk rot of maize

Causal organism: *Fusarium verticilliodes*

Nature and Recurrence of disease

Fusarium stalk rot is common in many of the corn-growing areas, it tends to be more severe in warm, dry regions. Fusarium species and other fungi can also cause seedling blight due of unimproved methods of seed production and carelessly handling, and planting techniques.

Symptoms

- Infected seedlings are usually stunted, and have pale green or purple leaves and poor roots.
- Symptoms of fusarium stalk rot in mature plants are difficult to distinguish from those of other stalk rots, but the internal tissues of affected stalks are usually reddish-brown and rotted.
- The discolouration may also be seen on the surface of the stalks near nodes. Stalks are weak and lodge easily.
- The pathogen commonly affects the roots crown regions and lower internodes.
- When split open, the stalk shows pink-purple dis-colouration.

Management

- For effective control of the disease, water stress at flowering should be avoided.
- Use balance dose of nutrients wherein potassium application helps in minimizing the disease.
- Use of bio-control agents (*Trichoderma* formulation) in furrows mixed with FYM @ 10g/kg at 10 days prior to its use in the field.
- It always advisable to practice crop rotation to minimize the disease incidence.

9. Downy mildews (DM)

Causal organism: *Peronosclerospora sacchari*

General comments

The member of the family "peronosporaceae" are obligate parasites and collectively referred as downy mildew fungi because of the specific symptoms produced in plants infected by them. In downy mildew diseases, the pathogen is seen s white, grey, brownish or purplish growth on the host surface. The superficial growth consisting of sporophores and spores is a tangled cottony or downy mildew.

Symptoms

- The main symptoms of downy mildew are legends developing on lower leaves as narrow chlorosis strips.
- Strips extend in parallel fashion, well defined margined delimited by veins.
- Downy whitish to creamy growth usually on the ventral surface of the infected leaves appears corresponding to stripes.

Management

- For control of downy mildew, the infected plants should be rogue out and destroyed.
- The planting of crop before onset of rains minimizes the incidence of mildew.
- Seed treatment with fungicides like Apron 35 WS @ 2.5 g/kg seed.
- Use resistant varieties.

Downy mildew

BAJRA

1. Green ear or downy mildew

Causal organism: ***Sclerospora graminicola***

Nature and Recurrence of Disease

This is a soil borne disease. Along with plant debris, the oospores survive in the soil for three to five years. On getting suitable moisture and other favourable conditions, the Oospores germinate by producing germ tubes, which infect the seedling.

Symptoms

- The leaves of infected plants show discoloration, yellowing and whitening.
- Under humid conditions, the leaves are covered with a downy mildew white growth of the fungus, which is prominent on the lower-surface.
- The leaves turn necrotic and are torn into shreds. The ears of the infected plants are transformed wholly or partly into green heads of small, twisted, leafy structures.

Management

- To prevent the secondary infection rogue out the diseased plants early in the season.
- Collect and destroy the diseased plant by burning.
- Seed treatment with Metalaxyl 35SD @3gm per kg of seed.
- Seed treatment with Carbendazim @2g/kg of seed
- 30-35 days after the sowing spray the crop with Ridomil MZ @0.25%

Green ear or downy mildew

2. Ergot

Causal organisms: *Claviceps fusiformis*

Nature and Recurrence of Disease

The disease is soil-borne. The conidia can survive for thirteen months, and this way, the fungus can survive from year to year by means of conidia.

Symptoms

- Caused by the fungus, at blossoming, a pinkish or light coloured fluid (honey dew) exudes from the spikelets on different parts of the ear.
- Later dark sticky patches appear on the ear. After fertilization, small dark sticky patches appear on the ear. After fertilization, small dark brown sclerotia appear in place of grains in the glumes. Seed set is poor or is completely inhibited.
- The ovary is replaced by a fungal mass with many folds on the surface. Ergot plants should not be fed to the cattle as it contains ergotoxine, which is quite harmful for cattle.

Management

- Before sowing dip the seed in 10% solution and remove the floating sclerotia.
- Treat the seed with carbendazim @2g/kg of seed.
- Ergot diseases can be managed by spraying the crop at boot leaf stage with Zineb @ 0.2%.

Ergot of bajra

3. Smut

Causal organism: *Tolysporium penicillariae*

Nature and Recurrence of Disease: This is soil borne disease as well as air borne disease. The spores germinate in soil producing promycelia and sporidium, and these sporidia are carried to the flowers of host pants by means of wind currents, where infection takes place.

Symptoms

- Individual grains in an ear get transformed into smut balls which burst open to release millions of spores which get disseminated and caused secondary infection on the portion of the ear which is enclosed by the sheath of the upper leaf.
- The intensity of the attack varies according to humanity in the areas.
- Only scattered grains (single or in group) in the ear are infected. The diseased grains are converted into deep brown to black spore mass of the fungus. The smut sori are oval or pear shaped.

Management

- Seed treatment with Carbendazim or Carboxin @2g/kg of seed.
- Remove the diseased ears in the season and destroy them.
- Before sowing treat the seed with an organo mercurial fungicide, viz. 2.5g of Cerasan/ Agrosan (phenyl mercury acetate) or 3g of Thiram/ Captan per kg of seed to prevent the introduction of smut into new areas.

Smut of bajra

SORGHUM

1. Anthracnose (foliar, head, root and stalk rot)

Causal organism: *Collectotricum graminicola, Glomerella graminicola* (Teleomorphs)

Symptoms

- *C. graminicola* is a fungal pathogen. The fungus can infect many grain crops such as barley, wheat sorghum and corn.
- The most common area of infection is the stalk *C. graminicola* produces three major symptoms types: leaf blight, stalk rot and top die-back.
- The leaf blight is characterized by round yellowing water soaked lesions on the leaves. These lesions usually occur early in the season and are how this pathogen is distinguished from other diseases.
- Top die back is the necrosis of the top leaves and stalk of the corn. This occurs around at the same time as grain formation.
- Blackening of the pith tissue in the stalk and also of the rind, beginning at the nodes closet to the soil. Along with these symptoms, seedling blight and post emergence off are also found.

Management

- Most effective methods of control is a one-year minimum of crop rotation to reduce anthracnose leaf blight.
- Use disease resistant varieties. Tillage systems that are able to fully bury corn residue deep underground along with one year crop rotation will reduce the source of inoculum greatly.

Anthracnose of sorghum

2. Charcoal rot

Causal organism: *Macrophomina phaseolina*

Symptoms

- The disease is well characterized by the presence of numerous black microsclerotia varying from 100 µm to 1mm in stems, leaves and roots and 50 µm-300µm in culture.
- It is an important pathogen of crops particularly where high temperatures and water stress occurs during the growing season.
- In soybean charcoal rot disease symptoms appear in mid summer during high temperatures and water stress occurs during the season.
- It appears in midsummer during high temperature and low soil moisture when the plant is stressed.
- Initial infections occur at seedling stage but remain latent until the soybean plant approaches maturity, Plants may wilt and die.
- Necrotic spots may also appear on the leaves of some plants due to a translocation toxin. Such toxin from culture and diseased plant have been shown to produce typical rot symptoms.

Management

- Seed treatment with carbendazim@ 2gm/kg of seed.
- Seed treatment with *Trichoderma* spp. @ 4gm/kg of seed.
- Seed treatment with *Pseudomonas fluorescens* @ 4gm/kg of seed.

Charcoal rot

PULSES

Moong/Mash

1. Cercospora leaf spot

Causal organism: ***Cercospora cruenta*** **and** ***Cercospora dolichii.***

Nature and Recurrence of disease

The disease may survive in the soil along with plant debris. The infection may take place by means of conidia found in the soil, in the following season.

Distribution: This is a wide spread disease, wherever this leguminous crop is grown.

Symptom

- Small, round or angular spots violet red in colour may be observed in leaves.
- These spots have grey coloured centre with radish purple border on the leaves stalk and pods.
- Affected pods become blackened.
- The conidiophores come out in clusters through the stomata. They are brown, septate and possess the knee joints marking the conidial scars, measuring 11 to 50 by 4 to 6µ. Conidia are 2 to 8 septate, measuring 50 to 100 by 4 to 5µ.

Management

- Spray Ziram or Dithane Z-78 or Dithane M -45 at the rate of 2.5g per litre of water.
- The first spray should be given with the appearance of first symptom of the disease. This should be followed by second spray after 10 days.

Cercospora leaf spot of urdbean

2. Bean blight

Causal organism: ***Ascochyta phaseolorum***

Nature and Recurrence of Disease

The disease is both externally and internally seed borne. The fungus survives in the seed in the form of mycelium as well as pycnidia. The secondary infection takes place by means of conidia.

Symptom

- Several spots develop on the leaves. They are innumerable, large brown in colour with a discoloured margin.
- Later on black minute dot like structures (fruiting bodies) appear on necrotic lesions which are arranged in circles.
- All the green parts of the plant attacked. Dark lesions appear on the stem and leaves first, then on the pods.
- On the stem, the leaves are oval or elongated, whereas on the leaves and pods, round and upto 1.25cm in diameter.
- When the lesions are fully developed, the margin is brown and the centre yellowish withered, and stunted with the minute, dark fructifications of the fungus, often concentrically arranged.

Management

- Spray the crop with Ziram or Dithane Z-78 or Mancozeb at the rate of 2.5g per litre of water.

Bean blight

3. Anthracnose

Causal organism: *Colletotrichum lindemuthianum*

General comment

Several species of genus cause anthracnose diseases in crops. In most cases, under favourable conditions, theses diseases cause heavy loss to production and producitvity of the crops.

Symptom

- Initially the disease is characterised by the production of dark brown circular spots on leaves. Later on the spots increase in size and develop in concentric ridges.
- All the aerial plant parts are infected.
- On seedlings, the cotyledons show small to large, brown to black sunken spots bearing pink spore masses. On growing plants leaves bear small angular and brown lesions adjacent to vein.
- The areas between the concentric ridges remain ash coloured. Dark coloured spots may also be seen on the pods.

Management

- Sow healthy seeds collected from disease free fields.
- Spray the Dithane Z-78 or Dithane M-45 @ 2.5g per litre of water.
- Use resistant varieties.

Anthracnose

4. Web blight

Causal organism: *Rhizoctonia solani*

Symptom

- It occurs on the foliage of the urd bean& moong bean. It starts from laminae or petioles or the young branches. Eventually, the top of plants become blighted and patches of such plants are conspicuously seen in the field.
- Whitish web-like growth develops on leaves in humid weather.
- Dark brown sclerotia develop on infected tissue. Infection on crop comes from the weeds in the field. Keeping the field weed-free helps to check the disease.
- The initial symptoms of the disease appear on the leaf lets as irregular greenish pale spreading spots with damp appearance. These spots often start from the apical portion of the leaf lets, increase rapidly in size to cover great parts of the leaf blade and then extend down to infect the petiole stem parts as well.
- The disease spreading from the affected to healthy parts causing typical web blight symptoms. The affected parts later turn dark brown, shrivel and dry out.

Management

- Drenching the soil with Carbendazim (50 wp) @ 0.1%.
- Use disease free seed.
- Prefer resistant varieties.

Web blight

5. Powdery mildew

Causal organism: ***Erysiphe polygoni***

Distribution

The disease is widely distributed all over the world. It is destructive disease in many parts of India, and appears in an epidemic form almost every year when the plants are in pod stage.

Nature and Recurrence of Disease

This is a soil-borne disease. The perithecia with their contaminated asci persist in the soil, where they reach from plant debris, until the following season, when they disintegrate on the host surface by producing germ-tube, and infect the crop. The spread of disease, generally takes place by means of conidia.

Symptom

- At the flowering time white powdery patches substances appear on the leaves and stem of the plants.
- The floury patches appear on both sides of the leaves. These may cover lagre areas of the plant.
- Later on the patches become dull coloured and are studded with black dot like cleistothesia.
- Resistance Varieties should be grown.

Powdery mildew

Management

- Spray sulfex or any wettable sulphur fungicides @ 2.5g per litre of water.
- Grow resistant variety.

6. Yellow mosaic disease

Causal organism: Mung bean yellow mosaic virus (MYMV) whitefly transmitted gemini virus.

Symptom

- The first symptom appears on the leaf lamina in the form of yellow, diffused, round scattered spots within a month after sowing.
- These spots expand rapidly and the leaves show yellow patches alternating with green colour of the leaves. The newly emerging leaves, after initial symptoms, show these symptoms from the very beginning.

Yellow mosaic disease

- The affected leaves later turn completely yellow and get reduced in size. The pods are smaller in size and the seeds are deformed.

Transmission: The virus is transmitted by insect vector and attempts to transmit mechanically were not successful.

Management

- Grow resistant varieties.
- Spray the crop with a mixture of 0.1% Metasystox+0.1% Malathion at 10 days interval to check the insect vector population. The spraying should be started with the first appearance of the disease.

7. Root rot

Causal organism: *Macrophomina phaseolina*

General comments

Root rots can also slow or stop plant growth and thus suppress plant quality. Diseased plants are usually smaller, less vigorous, produce fewer and/or small leaves, flowers, and pod than healthy plants of equal age. Flowering may be delayed when the plant's roots are rotted. As a result, the crop quality is very uneven. Root rots must be managed early in the disease if the losses are to be avoided.

Symptoms

- It produces dark lesions on leave, branches, stem and roots.
- The tissues of the affected portion become weak and easily shred.
- Pycinidia can be seen on the affected portion.

Root rot

Management

- Before sowing treat the seed with Captan of Thiram @3gm/kg of seed.
- Seed treatment with *Trichoderma viride*@4gm/kg of seed.
- Seed treatment with *Pseudomonas fluorescens* @ 4gm/kg of seed.

RAJMASH AND CLUSTER BEANS

1. Bean Rust

Causal organism: *Uromyces appendiculatus*

Nature and Recurrence of Disease: This is an air-borne disease.

Symptoms

- Initially, pale, raised spots appear on the upper leaf surface, corresponding to small white pustules on the lower surface. As summer progresses, tiny brown and then black pustules form on the lower leaf surface, often on the upper surface. The black pustules also form on the upper surface. Pods may also be affected.
- Heavily, infected leaves turn brown and die, and on severely affected plants may be stunted. It can also lead to shriveling and shedding of leaves.

Management

- Thin out dense growth on bean crops to help reduce humanity and so the risk of infection.
- Picking off affected leaves as soon as symptoms are seen should check the development of the disease. Symptoms tend on older leaves first.
- Dispose of (bin or burn) any severly affected plants as these will produce huge numbers of spores.
- Clear up and dispose of all debris (including fallen leaves) at the end of cropping.

Rust of bean

2. Bacterial wilt

Causal organism: *Corynebacterium flaccumfaceins*

Nature and Recurrence of disease

The pathogen can be seed transmitted or over winter in infected debris. Infections begin when the pathogen gains entrance to the vascular system through either seed infection or wounding in foliage and stems.

Symptoms

- Field symptoms consist of leaf wilting during periods of warm, dry weather or periods of moisture stress, wavy, intervenial, necrotic lesions surrounded by bright yellow borders. Appearance of yellow necrotic leaves lesions.
- Eventually a gradual systematic wilting of the plant persists, and the plant dies after turning straw colour. A vascular tissue inside the root and lower hypocotyls. Stem cankers and water-soaked pod also occur.

Management

- Use disease-free seed, and resistant varieties, practice crop rotation: rotating beans with other crops for two to three years, incorporating infected residues after the season, avoid re-using irrigation water.
- Seed may be treated with streptomycin, which will help to achieve better initial stands, but will not protect for the entire season. Consider using copper-based sprays during mid-vegetative or early flowering periods, depending upon weather conditions.

Bacterial blight of bean

3. Common blight

Causal organisms: *Xanthomonas campestris* pv. Phaseoli

Nature and Recurrence of Disease

The disease is both externally and internally seed borne. The fungus survives in the seed in the form of mycelium as well as pycnidia. The secondary infection takes place by means of conidia.

Symptoms

- On the leaves first appear as water-soaked spots. The spots enlarge into initially flaccid but then brown and necrotic lesions with lemon-yellow margins.
- Lesions often coalesce to cause such extensive tissue damage that defoliation occurs. Wilting is evident if the pathogen invades the plant vascular system.
- Symptoms on water-soaked spots that enlarge into dark red sunken lesions. Yellow exudates may form on leaves.

Management

- Clean and healthy seed should be sown.
- Hot water treatment may be given to reduce the internal infection.
- Seed should be obtained from disease free areas.
- Blighted plants should be pulled out by hands and burned.
- Resistant varieties should be grown.

Common blight

4. Halo blight

Causal organisms: *Pseudomonas syringae* pv. Phaseolicola

Symptoms

- On leaves, small, angular, water-soaked spots appear first on the lower leaf surface. As these spots increase in size a characteristic halo of yellow tissue develops around each water-soaked spot. The spots are 1/8 to ¼ in (3 to 6mm) in diameter. The halo may be upto 1 in (2.5cm) in diameter.
- On pods, the oval water-soaked spots may increase up to 3/8 in (9mm) in diameter and become slightly sunken and reddish-brown with age. Cream-coloured bacterial exudates are often found in pod lesions.
- Both leaf and pod lesions often coalesce. The upper foliage of diseased plants develop a characteristic yellow colour infected seed may be smaller than normal, have a wrinkled seed coat and be discussed.

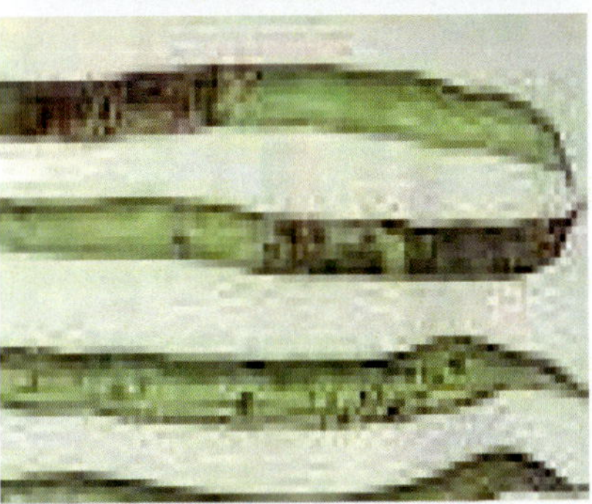

Halo blight in beans

Management

- Rotate crops, do deep ploughing and use pathogen-free seed.
- Treat seed with streptomycin.
- Spray diseased plants with copper-containing chemicals every 7 to 10 days after first observing symptoms.
- Plant resistant cultivars.
- Harvest before pod lesions turn brown.

5. Alternaria leaf spot

Causal organism: *Alternaria alternata*

Symptoms

- On the leaves, small, brown irregular-shaped lesions develop into large, gray-brown lesions with concentric rings.
- Leaf lesions do not always cross over major leaf veins: in such cases lesions may be angular in shape. When several lesions coalesce, a large portion of the leaf area become necrotic.

Alternaria leaf spot

- Sometimes the necrotic areas fall out and the leaf has shot hole appearance. Premature defoliation beginning with the lowest leaves may occur reddish-brown lesions that merge into long streaks develop on the pods from small water-soaked flecks.
- Old senescing leaves and pods are more susceptible than young leaves and pods.

Management

- Cultural management: Cultural measures are seldom warranted, but cultural measures such as wider plant and low spacings.
- Crop rotation.
- Chemical control may be needed.

6. Angular leaf spot

Causal organisms*: Phaeoisariopsiss griseola*

Symptoms

- Diseased plants are characterised by angular-shaped spots on the leaves. Initially such lesions are tannish-gray, but later they become dark brown or black. All aerial plant parts, including leaves, petioles, stems and pods can be infected, but symptoms are most recognizable on leaves. Lesions on leaves usually appear as brown spots with a tan or silvery centre that are initially confined to tissue between major veins, which gives it an angular appearance.
- Lesions can be observed on the underside of the leaf and appear slightly more pale than those on the upper surface of the leaflet A close look at lesions on the underside of a leaflet. As they increase in size, several may coalesce and proportions of the leaf area become infected and chloroti.

- During periods of high humidity, the undersurface of the leaves may have a black felt-like appearance due to formation of spores by the pathogen. Premature defoliation of the plant results. Diseased pods show circular-shaped spots with reddish-brown centres. Severely diseased plants have reduced vigor and poor yield.

Management

- Use cultural control practices such as a 2 years crop rotation.
- Planting pathogen free seed in well-drained soils, and deep ploughing to bury previously infected bean debris.
- Use chemical seed treatment and watch for development of new chemicals to protect bean foliage and pods.

Angular leaf spot

7. Anthracnose

Causal organisms: *Colletotrichum lindemuthianum*

Symptoms

- Symptoms are above ground parts of the plant appear as brick –red to dark brown lesions.
- On stems, leaf petioles and veins on the undersurface of leaves, lesions are usually sunken and elongate.
- On pods, lesions are sunken and circular.
- Infected seeds are usually discolored and may have sunken lesions.
- During periods of moist weather, gelatinous masses of pinkish spores may develop in infected areas. Severly diseased plants reduced in vigor and yield.

Management

- Pathogen free seed should be used.
- Always use disease resistant varieties.
- Plow under infected plant debris from fields soon after harvest and practice crop rotation (2 to 3 years).
- Watch for development of effective chemicals to control.

Anthracnose of beans

8. Powdery mildew

Causal organism: *Erysiphe poligoni*

General description: This is a fungal disease that affects a wide range of plants. Powdery mildew grows well in environments with high humidity and moderate temperatures. In an agricultural setting, the pathogen can be controlled using chemical methods, genetic resistance, and careful farming methods. It is important to be aware of powdery mildew and its management as the resulting disease can significantly reduce crop yields.

Symptoms

- All above –ground parts of the plant may be affected. The first symptoms are faint dark areas on the leaf that develop into small white powdery spots.
- It is one of the easier plant disease to identify, as its symptoms are quite distinctive.
- Infected plant display whit powdery spots on the leaves and stems. The lower leaves are the most affected, but the mildew can appear on any above ground part of the plant. These spots enlarge rapidly, coalesce and finally cover the entire leaf.

- If infection occurs early in the season, leaves may become dwarfed, turn yellow and fall off.
- On pods small, moist-looking circular spots develop into white powdery masses of pathogen mycelium and spores.

Management

- Use disease free seed.
- Use sulphur sprays or dusts or other chemicals early in the season before the pods become infected.

Powdery mildew

9. Rust

Causal organism: *Erysiphe poligoni*

Survival and spread

- The fungus survives mainly in alternate host.

Favourable conditions

- Rust progresses most rapidly in susceptible hybrids or varieties when the temperature is near 27°C with high humidity and frequent dews.

Symptoms

- Rust affects leaves and sometimes status, petioles and pods.
- The first symptoms appear on the undersurface of leaves as tiny, white, raised spots.
- These spots gradually enlarge and form reddish-brown pustules, which eventually erupt to release rusty masses of spores. Plant vigor may be severly diminished.

Management

- If available use resistant cultivars.
- Use crop rotation.
- Plough under old plant debris. Reduce plant densities.
- Chemical control with sulphur or other registered materials may be necessary.

Rust

10. Fusarium root rot

Causal organism: *Fusarium solani*

Survival and spread

- Causal organisms survive in soil and seed as resting mycelia or sclerotia for long period. These organisms often survive as saprophytes, living on dead plant material, or as dormant mycelia or spores.

Favourable conditions

- Diseases are prevalent under cool wet conditions that keep the soil temperatures below 13°C. *Rhizoctonia* root rot is most damaging when cool, wet conditions in the spring are followed by hot (25°C-29°C), dry conditions.

Symptoms

- Small, elongate, tan-red lesion in the lower hypocotyls and upper tap root. The lesions increase in number and size, often coalescing until the entire root system and lower hypocotyls show reddish brown necrosis.
- The rather dry lesions can penetrate root and hypocotyls tissue to develop deep elongate fissures leading to collapse under slight pressure. Diseased plants are stunted and unthrifty in relation to the severity of root rot. Severly diseased plants may die, or adventitious roots may form that help to keep the plant alive.

Management

- Deep dig compacted soils.
- Crops rotation should be done and avoid spread of infected plant debris and infested soil. Use properly fertilized and limed soils. Do not over irrigate. Use chemical for early season control. Use resistant cultivars.

Fusarium root rot

11. Pythium root rot

Causal organisms: *Pythium ultimum*

Nature and Recurrence of Disease

- It is a soil borne disease. These organisms often survive as saprophytes, living on dead plant material, or as dormant mycelia or spores and are prevalent under cool wet conditions that keep the soil temperatures below 13°C. *Rhizoctonia* root rot is most damaging when cool, wet conditions in the spring are followed by hot (25°C -29°C), dry conditions.

Symptoms

- Elongated water-soaked areas on hypocotyls and roots. These areas become slightly sunken, tannish-brown lesions which collapsed shrunken, tan-brown appearance because of the wet soft rot.
- Rot of both primary and secondary roots takes place; so much of the root system is destroyed.
- The plant is generally stunted or wilts and die.

Management

- Avoid over irrigation in early stages of crop development. Use resistant varieties
- Crop rotation with grain crops.
- Plant in well drained soils.
- Use wide spacing with plants.
- Chemical control.
- Determine root rot potential of bean fields to avoid those with high disease potential.

Pythium root rot

COWPEA

1. Leaf spot

Causal organism: ***Mycosphaerella cruenta***

Symptoms

On infected leaves (especially those more mature) look for brown or rust-coloured lesions that vary from circular to angular, are 2-10 mm, and may coalesce. Lesions may have a grey centre with a slightly reddish border. Lesions may dry and portions may fall out, giving the leaf a shot-hole appearance. Lesions and blemishes may occur on branches, stems and pods. Conidia develop at the centre on short conidiophores. Severely affected leaves become chlorotic.

Management

- Use resistant varieties and disease free seed.
- Deep ploughing with a mouldboard plough, which would bury residues.

Leaf spot of cowpea

2. Root rot

Anthracnose: ***Colletotrichum*** **spp.**

Symptoms

Tan to brown sunken lesions on leaves. Lesions merge to girdle stems and petioles. Lesions may become covered in pink spore masses during periods of wet weather.

Management

- The best method of controlling the fungus is to plant resistant varieties if available.
- Plant only certified disease-free seed. Practice good field sanitation such as removing crop debris from field after harvest to reduce levels of inoculums.

Root Rot

3. Ascochyta blight

Causal organism: ***Asochyta phaseolorum***

Symptoms

- Severe defoliation of plants; extensive lesions on stems and pods; if infection is severe then plants may be killed.

Management

- Plant disease-free seed.
- Applications of appropriate foliar fungicides, where available, may help to control the disease.

Ascochyta blight

4. Bacterial blight

Causal organism: ***Xanthomonas campestris***

Symptoms

- Water-soaked spots on leaves which enlarge and become necrotic. Spots may be surrounded by a zone of yellow discoloration. Lesions coalesce and give plant a burned appearance.

- Leaves that die remain attached to plant; circular, sunken, red-brown lesion may be present on pods. Pod lesions may ooze during humid conditions.

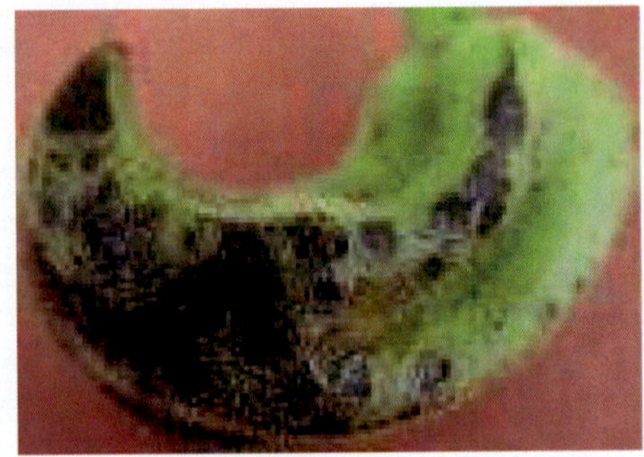

Bacterial blight of cowpea

Management

- Plant only certified seed; plant resistant varieties. Treat seeds with an appropriate antibiotic prior to planting to kill off bacteria.
- Spray plants with an appropriate protective copper based fungicide before appearance of symptoms.

5. Brown rust

Causal organism: *Uromyces* spp.

Symptoms

- Raised brown to black pustules on both sides of leaves; wilting plants; drying leaves dropping from plant

Management

- Spray of sulphur or potassium carbonate can help to control the disease.

Brown Rust

6. Cercospora leaf spot

Causal organism: *Cercospora canscens, Pseudocercospora cruenta*

Symptom

- Chlorotic spots on upper surfaces of leaves.
- Necrotic spots on leaves.

- Masses of spores on lesions which resemble black mats on lower leaf surface; defoliation of plants.
- Yellowing of leaves; circular, red lesions on leaves.

Management

- Remove all crop residue from field after harvest; plant disease-free seed.

Cercospora leaf spot

7. Fusarium wilt

Causal organism: *Fusarium oxysporum*

Symptoms

- Stunted plant growth; yellowing, necrotic basal leaves; brown-red or black streaks on roots that coalesce as they mature; lesions may spread above the soil line.

Management

- Control relies on cultural practices e.g. do not plant in same area more than once in any five year span or treating seeds with an appropriate fungicide prior to planting.

Fusarium wilt of cowpea

8. Yellow mosaic virus

Causal agent: Virus transmitted by White fly, a major insect pest attacking cowpea.

Symptoms

- First yellowish green colour spot appear on leaves.
- In severe case these spots converted into dark yellow in color. Reduction in yield.

Management

- Use resistant varieties.
- 20-25 days after sowing spray the crop with Metasystox @1.5ml/lt of water.
- Spray the crop with Malathion 50 EC @ 2ml/lt.
- Timely weeding and hoeing of crop.

Yellow mosaic of cowpea

OIL SEED

Sesame or Til

1. Phytophthora blight or Stem blight

Causal organism: *Phytophthora parasitica* var sesami.

Symptom

- The characteristic symptoms are the appearance of small water soaked irregular spots on the leaves. These spots later extend enlarge and cover the entire leaf area.
- The infected leaves become flaccid, drop and finally wither. Blackening of the stem at the soil level is the typical system of the disease.
- The affected capsules may also show whitish cottony growth of the fungus. Stem bark is loosened and can be peeled easily.
- In early infection plants fail to produce flower and die prematurely.

Management

- Spraying the crop with DithaneM-45 or DithaneZ-78 at the rate of 2.5 g per litre of water. It should be repeated after 10 days. (At least two times).

2. Alternaria leaf spot

Causal organism: ***Alternaria sesami***

Symptom

- The disease appears on leaves as irregular brown spots with concentric rings.
- Elongated lesions are seen on the stem and also the capsules get blighted.
- The plants are liked due to severe defoliation and stem infection.

Management

- Seed treated with Bavistin @ 2g per kg of seed before sowing.
- Spraying the crop with the DithaneZ-78 or DithaneM-45 on 40 -50th days of crop at the rate of 2.5g per litre of water.

3. Bacterial blight and Bacterial leaf spot

Causal organism: ***Xanthomonas sesami*** **(*Pseudomonas sesami*)**

Symptom

- Small soaked spots develop on the cotyledons of seedlings coming out from the infested seed.
- In older seedlings, spots of dark brown to black colour, round to irregular in shape appear on the true leaves.
- The spots may be coalesce to form bigger one at the later stage of development. Such leaves fall prematurely. Similar lesions may also be observed on the stem and petioles where the colour is black.

Management

- Seed treated with Streptocycline at 250 ppm/kg of seed for 30 minutes.
- Spraying the crop with Copper oxychloride (0.2%) + Streptomycin sulphate @ 1g/lt of water from 25 days of age of plants. 2-3 spraying will be necessary.

4. Phyllody

Causal organism: Mycoplasma like organism (MLOs). It is transmitted by Jassid, *Orosius abicinctus.*

Symptoms

- The disease becomes apparent at flowering stage of plant growth at about 45 days of sowing. One or floral parts are transformed fully or partially into green leafy structures followed by abundant vegetative growth.

- Heavily infected plants produce excessive abnormal growth with apical proliferation, small leaves, short internodes and abnormal branching.
- Top portion of the infected plants are changed into heavy bunches giving the branches a downward appearance. Ultimately plant look like witches broom or bunchy top.

Phyllody of til/sesame

Management

- Rogue out the diseased plants to prevent further spread of disease.
- Soil application of phorate (10G) @ 10kg \ ha at the time of sowing.
- Spraying the crop with Metasystox at the rate of 1ml per litre of water.

GROUNDNUT

1. Collar rot

Causal organism: ***Aspergillus niger***

Nature and Recurrence of Disease: This is a soil borne

Symptoms

- After sowing at any time seed can be affected, with it young plants collapsed and die soon after the emergence of disease.
- The hypocotly is attacked both by soil-borne *Aspergillus niger* and by the fungus growing from already infected cotyledons, or spore carrying on the seed surface which germinate after planting.
- One month of sowing disease damage the crop.

Management

- Seed treatment with Thiram @ 3g/kg.
- Seed treatment with Indofil M-45 @3gm / kg.
- Use resistant varieties.
- Avoid water stagnation in the field.
- Crop rotation after every 3 to 4 years.

Collar rot of groundnut

2. Botrytis blight

Causal organism: *Botrytis cinerea*

Symptoms

- Botrytis blight is most commonly seen at the end of the season when conditions are cool and wet.
- Symptoms appear first on vines or leaves that have been injured by tractor tires or freezing temperatures. Massive numbers of gray to brown spores can be produced on leaves and stems, giving them a fuzzy appearance.
- The stems and leaves have a water-soaked appearance at first, and then turn dark. Infrequently, numerous small spots are seen on the tops of leaves in the absence of other symptoms. These spots are light tan, irregular in shape, and have no obvious spores. Black sclerotia can be found on peanut pods.
- Although Botrytis blight does not usually cause serious losses, it can be alarming. Harvesting in a timely fashion and avoiding plant injury will reduce symptom incidence and severity.

Management

- Avoiding frost damage by planting early peanut varieties can help protect the plant from fungal colonization.
- Application of appropriate foliar fungicides (e.g. benomyl), where available, can help to control the disease.
- Use resistant varieties.
- Crop rotation after every 3 to 4 years to avoid disease.

Botrytis Blight

3. Charcoal rot

Causal organism: *Macrophomina phaseolina*

Symptoms

- Water soaked lesions on stems of seedlings close to soil line.
- Lesions girdle stem and kill seedlings.
- Lesions in similar area may be present in older plants.
- Lesions are initially water-soaked but turn brown, if lesions girdle the stem, plant wilts and branches die.
- Older plants are infected near the soil surface. The infection spreads upward within the stems and branches, and downward into the root system killing the plant. Sometimes, the disease is restricted to the roots. When pods are invaded their interior surfaces turn gray.

Management

- Rotating crop with rice for a period of 3-4 years can reduce the level of inoculum in the soil.
- Providing the plants with adequate irrigation and fertilization reduces susceptibility to the disease.
- There are currently no resistant varieties of peanut.
- Frequent irrigation to wet soil reduces the incidence of the disease.

Charcoal rot in groundnut

4. Cylindrocladium black rot

Causal organism: *Cylindrocladium crotalariae*

Symptoms

- Leaves on main stem turning chlorotic and wilting.
- Entire plant wilts very rapidly when there is a period of water stress following high moisture.
- Clusters of red-brown fungal bodies occur on stems, pegs and pods.
- Roots destroyed; roots blackened and shriveled.
- Reddish orange fungal bodies are formed in dense clusters on infected branches near the soil surface. Underground plant parts can be affected. In severe cases the disease may destroy pods and kernels.
- Diseased plants appear in the field in localized patches. Early symptoms are chlorosis and wilting of foliage on the main stem. Plants may not wilt but appear chlorotic and stunted. The disease may spread further in subsequent years.

Management

- The most effective method to control the disease is to plant peanut varieties that have some resistance to the disease.
- Rotation of crop with non host such as corn, cotton or tobacco may help to reduce inoculum in the soil.
- Application of appropriate soil fumigants in heavily infested fields can help to control the disease.
- Disease free seed should be used.

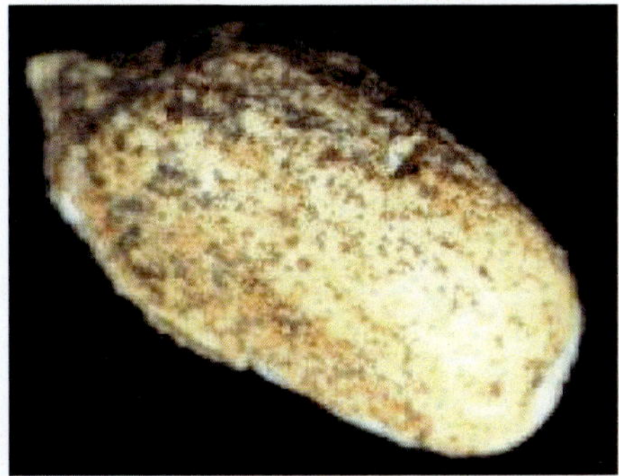

Black rot in groundnut

5. Circular leaf spot

Causal organism: ***Cercospora arachidicola*** **and** ***Cercospora personata.***

Nature and Recurrence of the Disease

This is a soil-borne disease. The primary infection is caused by means of conidia found on the plant debris in the soil. The spread of disease takes place by means of conidia which are dispersed by wind. The role of the perfect stage of the fungus is not clearly understood.

Symptoms

- Circular to irregular spots having reddish brown to dark brown color appear on leaf.
- Those spot are bigger in size as compared to that of late-leaf spots caused by *C. personata.*
- On both side of the leaf surface there is dark brown to black spot. There is large scale defoliation in groundnut due to these leaf spot.

Management

- Spray the crop with wettable sulphur (Sultef) 500-750 g in 200-300 lts. of water per acre.
- From the first week of August 3-4 times spray at fortnightly intervals.
- Spray the irrigated crop with Bavistin @2g per lt of water.
- Give three spray at fortnightly intervals, starting when the crop is 40 days old.
- Peanut crop debris should be plowed into soil after harvest and any volunteers removed from the non host crop.

Circular leaf spots

Section - C
Major Weeds of Rainfed Kharif Crops

3

Cereals

List of Weeds in *kharif* season crops

S. No.	Botanical Name	Common/local Name	Family
1.	*Abutilon indicum*	Indian mallow	Malvaceae
2.	*Acrachne racemosa*	Sat mooli	Poaceae
3.	*Acalypha indica*	Indian copperleaf	Euphorbiaceae
4.	*Acanthospermum hispidum*	Bristly starbur	Asteraceae
5.	*Amaranthus viridis*	Pigweed/Jungali chaulayi	Amaranthaceae
6.	*Amaranthus spinosus*	Spiny pigweed	Amaranthaceae
7.	*Bidens pilosa*	Sapnish blackjack	Asteraceae
8.	*Brachiaria reptans*	Running grass/ Para ghas	Poaceae
9.	*Celosia argentea*	Cock's ccmb	Amaranthaceae
10	*Chloris barbata*	Finger grass	Poaceae
11.	*Cleome viscosa*	Asian spider flower/Hulhul	Cleomaceae
12.	*Commelina benghalensis*	Day flower	Commelinaceae
13.	*Cucumis* spp.	Cucurbit weed	Cucurbitaceae
14.	*Cyperus rotundus*	Purple nut sedge/Deela	Cyperaceae
15.	*Cyperus esculentus*	Yellow nutsedge	Cyperaceae
16.	*Cynodon dactylon*	Bermuda grass	Poaceae
17.	*Dactyloctenium aegyptium*	Crowfoot grass/makra	Poaceae
18.	*Digera arvensis*	Kundra	Amaranthaceae
19.	*Digitaria sanguinalis*	Large crabgrass	Poaceae
20.	*Eragrostis* spp.	Lovegrass	Poaceae
21.	*Echinochloa colonum*	Chotta sawank	Poaceae
22.	*Eleusine indica*	Goose grass	Poaceae
24.	*Euphorbia hirta*	Petty spurge	Euphorbiaceae
25.	*Leucas aspera*	Thumbai	Lamiaceae
27.	*Parthenium hysterophorus*	Congress grass	Asteraceae
30.	*Phyllanthus niruri*	Stonebreaker/Hazardana	Phyllanthaceae
31.	*Physalis minima*	Rasbhari/chirpati	Solanaceae
32.	*Pluchea lanceolata*	Indian camphorweed	Asteraceae
33.	*Portulaca oleraceae*	Common purslane	Portulacaceae
34.	*Setaria glauca*	Yellow foxtail	Poaceae
35.	*Solanum nigrum*	Black nightshade	Solanaceae
36.	*Sorghum helepense*	Johnson grass	Poaceae

Contd.

S. No.	Botanical Name	Common/local Name	Family
37.	*Trianthema portulacastrum*	Horse purslane/Santhi	Aizoaceae
38.	*Tribulus terrestris*	Puncture vine	Zygophyllaceae
39.	*Tridax procumbens*	Coat buttons	Asteraceae

2. Weed Management techniques in cereals

Maize

Weeds emerge along with the germination of maize seeds. Major weeds emerge with the maize crop are *Trianthema portulacastrum* (Horse pursulane), *Echinochloa colonum* (Swank), *Digitaria sanguinalis* (Crab grass), *Commeline benghalensis* (Kaun makki), *Cyperus rotundus* (Purple nutsedge) etc. The critical period of crop-weed competition is upto 40-45 days after sowing (DAS). Reduction in maize yield occurs ranging from 25-65% under weedy situation. However, weeds management strategies limit the deleterious effects of weeds growing with the maize crop.

Management techniques

- Two hoeings to the crop should be given, one at 15 days and other at 30 days after sowing. Weeds within the rows can be effectively controlled by using traphali or 5 tinned hoes. This can be done with khurpa or hand blade hoe too.
- The crop should also be earthened up with bullock drawn ridger or with a spade when the crop is at knee high stage (after one month of sowing). It not only helps in uprooting weeds, but also helps in conserving rain water.
- Herbicide atrazine @ 0.75-1.0 kg/ha in 600-800 L of water should be sprayed on soil surface during 1-2 days after sowing as pre-emergence.
- If in any case farmer misses to apply pre-emergence herbicide in maize, post emergence application of 2, 4-D Na salt @ 1.0 kg/ha for broad-leaved weeds, tembotrione @ 100-120 g/ha for grassy and sedges and halosulfuron-methyl @ 67.5 g/ha for sedges and grassy weeds at 2 or 3rd leaf stage (15-20 DAS) of weeds in maize can be helpful in minimizing the weed density.
- For intercropping systems, atrazine should not be applied. In maize + pulse intercropping system, pre-emergence application of pendimethalin @ 1.0 kg/ha is to be used.
- In maize + soybean intercropping system, pre-emergence application of alachlor @ 2.0 kg/ha followed by one hand weeding is to be practiced.

- In case of pre-mergence herbicide application, spraying should be done within 2 days after sowing ensuring adequate soil moisture at the time of spray.
- The soil should not be disturbed during or immediately after application pre-emergence herbicides.
- The mulching of crop residue if available is to be used for reducing weed population besides conservation of soil moisture.
- No intercultural operations should be done after 6 weeks of sowing as it would lead to pruning of fine roots and finally reduce the production.

Pearl millet/Sorghum

Major weeds infesting the field are *Echinochloa colonum* (Swank), *Digitaria sanguinalis* (Crab grass), *Trianthema portulacastrum* (Horse purslane), *Cyperus rotundus* (Purple nutsedge), *Solanum nigrum* (Makoi) etc.

Management techniques

- Keep sorghum field free of weeds from second week after germination to till 5th week.
- If sufficient moisture is available, spray atrazine @ 0.75 - 1.0 kg/ha as pre-emergence (1-3 DAS) followed by 2,4-D @ 0.5-1.0 kg/ha at 20-25 DAS using 500-600 L of water/ha.
- In case, if herbicides are not used, hand weeding twice on 10-15 DAS and 30-35 DAS. If pulse crop is to be raised as an intercrop with sorghum, do not use atrazine, instead spray pendimethalin @ 0.75 kg/ha as pre-emergence.

Ragi

Ragi crop is also infested with almost similar weed species as pearl millet.

Management techniques

- Apply oxyfluorfen @ 50 g/ha as pre-emergence at 1-3 DAS using 500-600 L of water/ha followed by one hand weeding on 20 DAS. Herbicides application should be done when there is sufficient moisture in the soil.
- For broad leaved weeds control, apply 2,4-D Ethyl ester or 2,4-D Na salt @ 0.5 kg/ha as post-emergence herbicide at 10-15 DAS depending on the moisture availability.
- If pre-emergence herbicide is not applied, hand weeding twice at 15-20 and 30-35 DAS should be done to keep the weed population under control.

3. Weed management techniques in Pulses

Greengram/Blackgram

Initial 25-30 days after sowing is the critical period of crop-weed competition. Weeds dominating the moongbean and urdbean crops in field are *Cynadon dactylon* (Bermuda grass), *Cyperus rotundus* (Purple nutsedge) *Echinochloa colonum* (Swank), *Digitaria sanguinalis* (Crab grass), etc. During this period, crop should be kept with minimum competition from weeds. Due to continuous rains in *kharif* crops suffers from intense weed competition.

Management techniques

- Pre-emergence application of pendimethalin @ 0.75-1.0 kg/ha or pendimethalin + imazethapyr @ 1000 g/ha in 600-800 L of water followed by one hand weeding at 20-25 DAS is found effective in managing weeds in greengram / blackgram.
- If in any case farmer miss to apply pre-emergence herbicide, post-emergence application of quizalofop-ethyl 40-50g/ha or imazethapyr @ 70 g/ha or imazamox + imazethapyr @ 70 g/ha at 15 –20 DAS has been found effective and economical.
- However, two hand weedings at 15 and 30 days after sowing have been found to be best cultural practice in managing the weeds in greengram / blackgram.

Pigeonpea

Major weeds dominating the pigeonpea fields are *Cynadon dactylon* (Bermuda grass), *Cyperus rotundus* (Purple nutsedge) *Echinochloa colonum* (Swank), *Digitaria sanguinalis* (Crab grass), etc.

Management techniques

- For pigeonpea, the first 45-60 days appear critical period for weed competition, although this period may vary with the genotype and time of sowing.
- In pigeonpea pre-emergence application of pendimethalin @ 0.75-1.0 kg/ha or oxadiazon @ 0.75-1.0 kg/ha or alachlor @ 1.0-1.5 kg/ha or metolachlor @ 1.0-1.5 kg/ha followed by directed application of paraquat @ 0.4 kg/ha at 6-8 WAS or directed application of glyphosate 1.0 kg/ha at 45 DAS or integrated with one hand weeding at 50 DAS or with one intercultural operation is found effective in controlling weeds at satisfactory level.

- Weeds in pigeonpea can also be controlled by post-emergence application of imazethapyr @ 70-100 g/ha at 15-20 DAS *fb* hand weeding at 50 DAS or pendimethalin @ 0.75-1.0 kg/ha *fb* quizalofop-ethyl at 15-20 DAS *fb* hand weeding at 50 DAS.
- Intercropping also reduces weed infestation by 50 to 70 percent. Intercropping of maize and sorghum can suppress weeds for longer period. With short duration pigeonpea, low stature crops such as cowpea, greengram, blackgram, groundnut and soybean as smother crops can minimize the weed problem.

Cowpea

Major weeds infesting cowpea in field are *Cynadon dactylon* (Bermuda grass), *Cyperus rotundus* (Purple nutsedge) *Echinochloa colonum* (Swank), *Digitaria sanguinalis* (Crab grass), etc.

Management techniques

- Pendiethalin @ 0.75 kg/ha as pre emergence at 1-2 DAS should be applied to manage the initial infestation of weeds.
- Post emergence application of imazethapyr @ 70 g/ha or fenoxaprop-p-ethyl @ 75 g/ha or quizalofop @ 40 g/ha at 15-20 DAS using 500 L of water for spraying one ha to manage weeds in cowpea. After this, one hand weeding on 30-35 DAS gives weed free environment throughout the crop period.
- If herbicides are not applied, give two hand weedings on 15 and 30 days after sowing.

Cluster bean

Cynadon dactylon (Bermuda grass), *Cyperus rotundus* (Purple nutsedge) *Echinochloa colonum* (Swank), *Digitaria sanguinalis* (Crab grass), etc are major weeds infesting the field.

Management techniques

- Critical period for crop-weed competition in cluster bean has been identified as 30-35 DAS and presence of weeds beyond these results in yield reductions by 47 to 92%.
- Pre-emergence application of pendimethalin @ 1.0 kg/ha (PE) + one hand weeding at 30-35 DAS is required to manage weeds in cluster bean.
- Post emergence application of imazethapyr @ 60 g/ha at 20 days after sowing (DAS) is also found effective in managing weed emerge after germination of crop plants.

- In cluster bean imazethapyr + imazamox (ready mix) @ 40 g/ha applied as post-emergence at 3-4 leaf stage (20 DAS) also found suitable for weed control.

Rajmash

Management techniques

- Pre-emergence application of pendimethalin @ 1.00 kg/ha reduces the weed population during the initial establishment of crop.
- Post emergence application of imazthapyr @ 70 g/ha or quizalofop-ethyl 60 @ g/ha as post-emergence at 20 DAS found suitable for weed management in rajmash.
- Under sufficient availability of labour farmer can go for hoeing/hand weedings at 15 and 30 DAS to control weeds.

4. Weed management in Oilseeds

Sesame Til

Major weeds infesting til crop are *Cynadon dactylon* (Bermuda grass), *Cyperus rotundus* (Purple nutsedge) *Echinochloa colonum* (Swank), *Digitaria sanguinalis* (Crab grass), etc.

Management techniques

- Sesame is sensitive to weed competition during the first 25-35 DAS.
- A minimum of two weedings / inter-cultivations, one after 15 DAS and another at 35 DAS are required to keep the field relatively weeds free. Row seeded crop facilitates use of blade harrows for inter-cultivation.
- Herbicides use, especially under rainfed conditions, is very limited due to low yield, which may not compensate for the cost of herbicides. If necessary, alachlor @ 1.0 kg/ha or thiobencarb @ 2.0 kg/ha can be used as pre-emergence spray for effective control of weeds.
- Use of pre-emergence herbicides followed by one hand weeding around 30 DAS is the most appropriate way of weed management in sesame.

Major weeds in *kharif* season crops

Abutilon indicum

Acrachne racemosa

Acalypha indica

Acanthospermum hispidum

Amaranthus spp.

Bidens pilosa

Brachiaria reptans

Celosia argentia

Cleome viscosa

Commelina benghalensis

Cucumis spp

Cyperus rotundus

Cynodon dactylon

Dactyloctenium aegyptium

Digera arvensis

Digitaria sanguinalis

Echinochloa colonum

Eleusine indica

Euphorbia hirta

Leucas aspera

Phyllanthus niruri

Physalis minima

Pluchea lanceolata

Portulaca oleraceae

Setaria glauca

Solanum nigrum

Sorghum helepense

Trianthema portulacastrum

Tribulus terrestris

Tridax procumbens

5. Non-crop lands weeds and their management

1. Parthenium hysterophorus

Parthenium hysterophorus is an aggressive ubiquitous annual herbaceous weed with no economic importance. It is commonly known as carrot grass, bitter weed, star weed white top etc. *Parthenium* can be managed by following methods:

Parthenium hysterophorus at seedling stage

Parthenium hysterophorus at flowering plant

1. Physical pulling/uprooting of *Parthenium* plants by covering hands with gloves or polythene bags in rainy season.
2. Spraying 10-15% common salt (brine solution) followed by controlled burning
3. Herbicides
    a. 2, 4-D sodium, amine or ester @ 0.8-1.0 kg/ha
    b. Gylphosate 1-1.5%
    c. 2, 4-D @ 1 kg/ha + paraquat @ 0.5 kg/ha
4. Metribuzin 0.5%
5. Seeding with *Cassia* spp., *Mirabilis jalapa* and planting of marigold may be done wherever possible for providing alternate vegetation that suppress *Parthenium.*
6. *Parthenium* compost making should be encouraged by utilizing this weed before flowering for providing nutrient rich manure to the crops.

2. Saccharum spontaneum

It is a perennial grass, growing up to three meters in height, with spreading rhizomatous roots. Leaves are harsh and linear, 0.5 to 1 meter long; 6 to 15 mm wide. *Saccharum spontaneum* can be managed by following methods:

Saccharum spontaneum at seedling stage

Saccharum spontaneum at flowering plant

1. Cut *Saccharum* plants just above the ground level during December-January.
2. Spray 1 per cent glyphosate solution during May-June on the regenerated fresh growth.
3. Plant either hybrid napier or cenchurs slips during raining season (July-August) to provide a grass cover to the treated area to avoid further encroachment by the weed.
4. Spot treatment of the herbicide on the regenerated growth, if any, may be given in October-November followed by gap filling by either of the above grasses during January-February.

Lantana camara flowering *plant*

Lantana camara seed

3. *Lantana camara*

Lantana camara is a highly variable ornamental shrub; it is generally deleterious to biodiversity and human activities.

The management of *Lantana camara* can be done by cutting the bushes near the ground level in the month of April followed by herbicidal spray (Glphyosate 1 per cent) on about 30 cm regenerated growth of *Lantana camara* during June and subsequent planting of fodders like *Setaria* or Napier to provide cover to the treated land so as to avoid further reoccurrence of the lantana bushes.

General voluntary vegetation control

1. Glyphosate 1.0-1.5% + urea (2%)
2. Paraquat 1kg/ha
3. Dalapon 5-6 kg/ha

Section - D
Nutritional Deficiency Disorders of Rainfed Kharif Crops

4

Essential Plant Nutrients

Proper nutrition is essential for satisfactory crop growth and production. The use of soil tests can help to determine the status of plant available nutrients to develop fertilizer recommendations for achieving optimum crop production. The profit potential for farmers depends on producing enough crop per acre to keep production costs below the selling price. Efficient application of the correct types and amounts of fertilizers for the supply of the nutrients is an important part of achieving profitable yields.

There are at least 17 elements known to be essential for plant growth. Carbon (C), Hydrogen (H), and Oxygen (O) are derived from carbon dioxide (CO_2) and water (H_2O). Nitrogen (N), Phosphorus (P), Potassium (K), Sulphur (S), Calcium (Ca), Magnesium (Mg), Boron (B), Chlorine (Cl), Copper (Cu), Iron (Fe), Manganese (Mn), Molybdenum (Mo) and Zinc (Zn) are normally derived from the soil in the form of inorganic salts. 94 to 99.5 per cent of fresh plant material is made up of carbon, hydrogen and oxygen. The other nutrients make up the remaining 0.5 to 6.0 per cent.

Macronutrients

Refer to those elements that are used in relatively large amounts, whereas micronutrients refer to those elements that are required in relatively small amounts (Table 1).

Table 1: Essential Plant Nutrients

Supplied from air and water	Supplied from soil and fertilizer sources	
	Macronutrients	Micronutrients
Carbon (C) Hydrogen (H) Oxygen (O)	Nitrogen (N) Phosphorous (P) Potassium (K) Sulphur (S) Calcium (Ca) Magnesium (Mg)	Zinc (Zn) Copper (Cu) Iron (Fe) Manganese (Mn) Boron (B) Chlorine (Cl) Molybdenum (Mo) Cobalt (Co), Nickel (Ni) Silicon (Si)

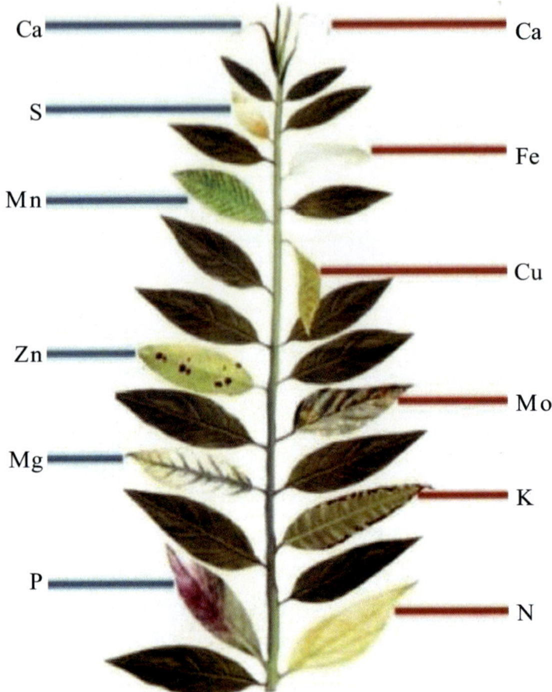

Plant Nutrient Deficiency Chart

Nutrient deficiency symptoms are not commonly found in modern crops under good agriculture production practices. The classic signs of major nutrient deficiencies (or toxicities) do not normally appear in well-managed fields. However, knowledge of those symptoms and the conditions that cause them can be an important diagnostic tool. In general, if deficiency symptoms do appear on crops during the growing season, significant yield loss has already occurred. Crops are more likely to suffer from "hidden hunger" or conditions which tend to limit yields or quality without apparent symptoms. More than one nutrient deficiency may occur at the same time. Nutrient deficiencies can be prevented with intensive soil testing programs, followed by plant tissue testing. Tissue testing is an under-used tool that could help identify problems in time to take corrective action during the growing season.

1. Nitrogen

Nitrogen (N) is the kingpin among the macronutrients and is the most commonly applied fertilizer nutrient to crops. Nitrogen is a major component of proteins, hormones, chlorophyll, vitamins and enzymes essential for plant life. Nitrogen metabolism is a major factor in stem and leaf growth (vegetative growth). While the excess quantity can delay flowering and fruiting. Deficiencies can reduce yields, cause yellowing of the leaves and stunt growth.

Functions

- Promotes growth of leaves and stems.
- Gives dark green colour and improves quality of foliage.
- Necessary to develop cell proteins and chlorophyll.
- Necessary for formation of amino acids, the building blocks of protein.
- Essential for plant cell division, vital for plant growth.
- Directly involved in photosynthesis.
- Necessary component of vitamins.
- Aids in production and use of carbohydrates.
- Affects energy reactions in the plant.

Deficiency symptoms

- Deficiency noted when leaves are a sick, yellow-green colour.
- Short stems, small leaves, pale coloured leaves and flowers.
- Slow and dwarfed plant growth.

Nitrogen deficiency symptoms in maize

2. Phosphorus

Phosphorus is the second most commonly applied fertilizer nutrient. The plant takes up P as inorganic anion ($H_2PO_4^-$ and HPO_4^{2-}). The form of uptake is largely governed by pH. The plant roots contain more of active sites for absorption of primary orthophosphate than the secondary orthophosphate. At low pH, uptake of HPO_4^{2-} is more than HPO_4^{2-}. At pH 7, both forms are taken up equally. Low pH (<4) results in phosphate being chemically locked up in organic soils. Phosphorus must be applied close to the plant's roots to facilitate better utilization by plant. Large applications of phosphorus without adequate levels of zinc can cause a zinc deficiency.

Functions

- Stimulates early formation and growth of roots.
- Provides fast and vigorous growth.
- Essential for flower and fruit formation.
- Stimulates flowering and seed development.
- Involved in photosynthesis, respiration, energy storage and transfer, cell division and enlargement.
- Improves quality of fruits, vegetables, and grains.
- Vital to seed formation.
- Helps plants survive harsh winter conditions.
- Increases water-use efficiency.
- Hastens maturity.

Deficiency symptoms

- Deficiency symptoms are purple stems and leaves; plant maturity and growth are retarded.
- Yields of fruit and flowers are poor.
- Premature drop of fruits and flowers may often occur.
- Deficiency noted when leaves are a sick, yellow-green colour.
- Short stems, small leaves, pale coloured leaves and flowers.
- Slow and dwarfed plant growth.

Phosphorus deficiency symptoms in maize

3. Potassium

Potassium remains in an ionic form in the plant cells and tissues and plays an important role in osmoregulation. K plays an important role in fibre development and the turgor driven expansion of fibre cells ultimately determines the fibre length. Equally important is its role in enzyme activation. Potassium is necessary for formation of sugars, starches, carbohydrates, protein synthesis and cell division in roots and other parts of the plant. It helps to adjust water balance, improves stem rigidity and cold hardiness, enhances flavour and colour of fruit and vegetable crops, increases the oil content of fruits and is important for leafy crops.

Functions

- Used to form carbohydrates and proteins.
- Formation and transfer of starches, sugars, and oils.
- Increases disease resistance, vigour and hardiness.
- Carbohydrate metabolism and the break down and translocation of starches.
- Increases photosynthesis.
- Increases water-use efficiency.
- Essential to protein synthesis.
- Important in fruit formation.
- Activates enzymes and controls their reaction rates.
- Improves quality of seeds and fruit.
- Improves winter hardiness.
- Increases disease resistance.

Deficiency symptoms

- Deficiency symptoms include mottled, spotted, streaked or curled leaves.
- Scorched, burned, dead leaf tips and margins.

Potassium deficiency symptoms in maize

Complete –N –P –K –N,P,K

Macronutrient deficiencies in beans

4. Calcium

Calcium is the second of the secondary nutrients taken up by the plants as Ca^{2+} ion. It is readily transported to the root surfaces by mass flow. Inspite of sufficient amounts of Ca^{2+} plant uptake is limited because of its absorption by young root tips. This is the reason why Ca^{2+} uptake is less compared to K. Calcium activates enzymes, is a structural component of cell walls, influences water movement in cells and is necessary for cell growth and division. Some plants must have calcium to take up nitrogen and other minerals. Calcium is easily leached. Calcium, once deposited in plant tissue, is immobile (non-translocatable) so there must be a constant supply for growth.

Functions

- Improves plant vigour.
- Influences intake and synthesis of other plant nutrients.
- Important part of cell walls.
- Utilized for continuous cell division and formation.
- Involved in nitrogen metabolism.
- Reduces plant respiration.
- Aids translocation of photosynthesis from leaves to fruiting organs.
- Increases fruit set.
- Essential for nut development in peanuts.
- Stimulates microbial activity.

Deficiency symptoms

- Deficiency causes stunting of new growth in stems, flowers and roots.
- Symptoms range from distorted new growth to black spots on leaves and fruit.
- Yellow leaf margins may also appear.
- Symptoms of deficiency include small developing leaves, wrinkled older leaves.
- Dead stem tips.

5. Magnesium

Magnesium is a critical structural component of the chlorophyll molecule and is necessary for functioning of plant enzymes to produce carbohydrates, sugars and fats. It is used for fruit and nut formation and is found essential for germination of seeds. The uptake of Mg is similar to that of Ca. Due to cationic competitive effects, uptake of Mg is lower than K or Ca. The most important function of Mg is its occurence in the centre of the chlorophyll molecule. The other major role Mg^{2+} is as a cofactor in enzymes activating photophosphorylation processes. Inadequate levels of Mg in the plant can inhibit CO_2 assimilation. It is also required to activate the enzyme ribulose diphosphate carboxylase. It plays an important role in N metabolism. Proportion of protein-N decreases in Mg deficient crop caused by the dissociation of the ribosomes. Mg helps in translocation of cellulose and determines fibre quality. Deficient plants appear chlorotic, show yellowing between veins of older leaves; leaves may droop. It can be applied as a foliar spray to correct deficiencies.

Functions

- Influences the intake of other essential nutrients.
- Key element of chlorophyll production.
- Improves utilization and mobility of phosphorus.
- Activator and component of many plant enzymes.
- Directly related to grass tetany.
- Increases iron utilization in plants.
- Influences earliness and uniformity of maturity.
- Assists in translocation of phosphorus and fats.

Magnesium deficiency symptoms in wheat

Deficiency symptoms

- Deficiency symptoms include interveinal chlorosis-yellowing of leaves between green veins.
- Leaf tips curl or cup upward.
- Slender, weak stems.

6. Sulfur

Sulfur is a structural component of amino acids, proteins, vitamins and enzymes and is essential to produce chlorophyll. It imparts flavour to many vegetables. Deficiency shows as light green leaves. Sulfur is readily lost by leaching from soils and should be applied with a nutrient formula. Some water supplies may contain sulfur.

Functions

- Promotes root growth and vigorous vegetative growth.
- Essential to protein formation.
- Integral part of amino acids.
- Helps develop enzymes and vitamins.
- Promotes nodule formation on legumes.
- Aids in seed production.
- Necessary in chlorophyll formation (though it isn't one of the constituents).

Deficiency symptoms

- The young leaves are light green with lighter coloured veins.
- Yellow coloured leaves and stunted growth.

Sulphur deficiency symptoms in maize

Healthy leaves shine with a rich dark green colour when adequately fed.

Phosphate: deficiency turns leaf margins reddish purple, particularly on young plants.

Potash: deficiency appears as a fring or drying along the tips and edges of lower leaves

Nitrogen: hunger sign is yellowing that starts at the tip and moves along middle of leaf.

Magnesium: deficiency causes whitish strips along the veins and often a purlish colour on the undersides of the lower leaves.

Nutrient deficiency systems in maize

Micronutrients

1. Iron

Iron is necessary for many enzymatic functions and as a catalyst for the synthesis of chlorophyll. It is essential for the young growing parts of plants. Deficiencies are pale leaf colour of young leaves followed by yellowing of leaves and large veins. Iron is lost by leaching and is held in the lower portions of the soil structure. When soils are alkaline, iron may be abundant but unavailable. Applications of an acid nutrient formula containing iron chelates held in soluble form, should correct the problem.

Functions

- Essential for chlorophyll production.
- Helps carry electrons to mix oxygen with other elements.
- Promotes formation of chlorophyll.
- Acts as an oxygen carrier.
- Regulates reactions involving cell division and growth.

Deficiency symptoms

- Deficiency symptoms include mottled and interveinal chlorosis in young leaves.
- Stunted growth and slender, short stems.

2. Zinc

Zinc is a component of enzymes or a functional cofactor of a large number of enzymes including auxins (plant growth hormones). It is essential to carbohydrate metabolism, protein synthesis and internodal elongation (stem growth). Deficient plants have mottled leaves with irregular chlorotic areas. Zinc deficiency leads to iron deficiency causing similar symptoms. Deficiency occurs on eroded soils and is least available at a pH range of 5.5 - 7.0. Lowering the pH can increase zinc availability to the point of toxicity.

Functions

- Helps plant metabolism function.
- Aids in reproduction.
- Aids plant growth hormones and enzyme system.
- Necessary for chlorophyll production.
- Necessary for carbohydrate and starch formation.
- Aids in seed formation.

- Deficiency includes retarded growth between nodes (rosetted).
- New leaves are thick and small.
- Spotted between veins, discoloured veins.

Causes

- Under application of Zn fertilization.
- Soil has high pH (>7.5).

3. Boron

Boron does not easily move around the plant therefore a deficiency is most likely to be seen in the younger tissues first. Boron is necessary for cell wall formation, membrane integrity, calcium uptake and may aid in the translocation of sugars. Boron affects at least 16 functions in plants. These functions include flowering, pollen germination, fruiting, cell division, water relationships and the movement of hormones. Boron must be available throughout the life of the plant. It is not translocated and is easily leached from soils. Deficiencies kill terminal buds leaving a rosette effect on the plant. Leaves are thick, curled and brittle. Fruits, tubers and roots are discoloured, cracked and flecked with brown spots.

Functions

- Affects water absorption by roots.
- Translocation of sugars.
- Essential for germination of pollen grains and growth of pollen tubes.
- Essential for seed and cell wall formation.
- Promotes maturity.
- Necessary for sugar translocation.
- Affects nitrogen and carbohydrate.

Deficiency symptoms

- Deficiency symptoms include short, thick stem tips.
- Young leaves of terminal buds are light green at base.
- Leaves become twisted and die.

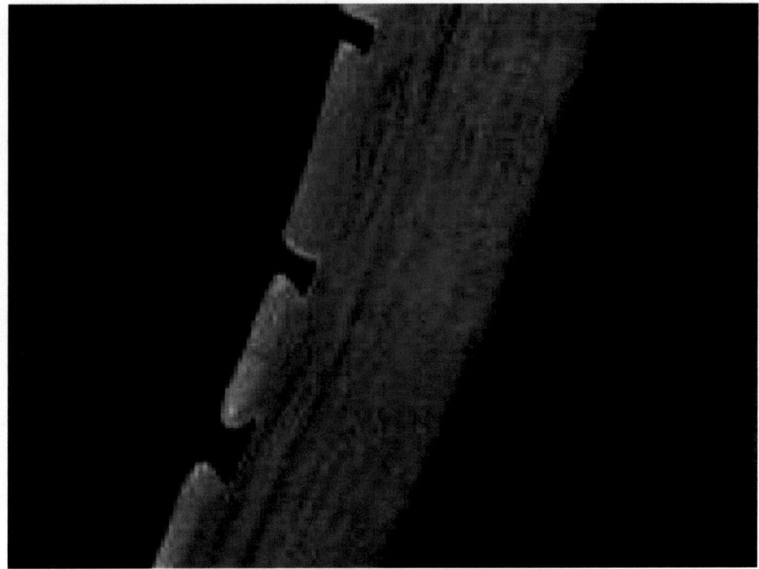

Saw tooth notches on wheat leaf

4. Manganese

Manganese is involved in enzymatic activities of photosynthesis, respiration, and nitrogen metabolism. Deficiency in young leaves may show a network of green veins on a light green background similar to an iron deficiency. In the advanced stages, the light green parts become white and leaves are shed. Brownish, black or grayish spots may appear next to the veins. In neutral or alkaline soils plants often show deficiency symptoms. In highly acidic soils, manganese may be available to the extent that it results in toxicity.

Functions

- Aids in plant metabolism.
- Helps in nitrogen transformation.
- Functions as a part of certain enzyme systems.
- Aids in chlorophyll synthesis.
- Increases the availability of P and Ca.

Deficiency symptoms

- Deficiency symptoms include interveinal chlorosis.
- Young leaves die.

Manganese deficiency symptoms in maize

5. Molybdenum

Molybdenum is a structural component of the enzyme that reduces nitrates to ammonia. Without it, the synthesis of proteins is blocked and plant growth ceases. Root nodule (nitrogen fixing) bacteria also require it. Seeds may not form completely and nitrogen deficiency may occur if plants lack molybdenum. Deficiency signs are pale green leaves with rolled or cupped margins.

Functions

- Aids in plant development and reproduction.
- Required to form the enzyme "nitrate reductase" which reduces nitrates to ammonium in plant.
- Aids in the formation of legume nodules.
- Needed to convert inorganic phosphates to organic forms in the plant.

Molybdenum deficiency symptoms in maize

Deficiency symptoms

- Deficiency symptoms include stunted growth.
- Yellow leaves, upward curling leaves.
- Leaf margin burn.

6. Chlorine

Chlorine is involved in osmosis (movement of water or solutes in cells), the ionic balance necessary for plants to take up mineral elements and in photosynthesis. Deficiency symptoms include wilting, stubby roots, chlorosis (yellowing) and bronzing. Chloride, the ionic form of chlorine used by plants, is usually found in soluble forms and is lost by leaching. Some plants may show signs of toxicity if levels are too high.

7. Nickel

Nickel has just recently won the status as an essential trace element for plants according to the Agricultural Research Service Plant, Soil and Nutrition Laboratory in Ithaca, NY. It is required for the enzyme urease to break down urea to liberate the nitrogen into a usable form for plants. Nickel is required for iron absorption. Seeds need nickel in order to germinate. Plants grown without additional nickel will gradually reach a deficient level at about the time they mature and begin reproductive growth. If nickel is deficient plants may fail to produce viable seeds.

8.Sodium

Sodium is involved in osmotic (water movement) and ionic balance in plants.

9. Cobalt

Cobalt is required for nitrogen fixation in legumes and in root nodules of non-legumes. The demand for cobalt is much higher for nitrogen fixation than for ammonium nutrition. Deficient levels could result in nitrogen deficiency symptoms.

10. Silicon

Silicon is found as a component of cell walls. Plants with supplies of soluble silicon produce stronger, tougher cell walls making them a mechanical barrier to piercing and sucking insects. This significantly enhances plant heat and drought tolerance. Foliar sprays of silicon have also shown benefits reducing populations of aphids on field crops. Tests have also found that silicon can be deposited by the plants at the site of infection by fungus to combat the penetration of the cell walls by the attacking fungus. Improved leaf erectness, stem strength and prevention or depression of iron and manganese toxicity have all been noted as effects from silicon. Silicon has not been determined essential for all plants but may be beneficial for many.

References

Anonymous, University of Nebraska-Lincoln, Institute of Agriculture and Natural Resources, Crop Watch. Bacterial wilt by Robert Harveson, Extension Plant Pathologist.

Atwal, A.S. and Dhaliwal, G.S. 1997. Agricultural Pests of South Asia and their Management. Kalyani Publishers. Ludhiana, India.

Aulakh, M.S., Sidhu, B.S., Arora, B.R. and Singh, B. 1985. Content and uptake of nutrients by pulses and oilseed crops. *Indian Journal of Ecology*. **12:**238-242.

B.P. Panday. 1994. A Text Book of Plant Pathology. Publisher S. Chand and Company Limited, Ram Nagar, New Delhi.

Goud, V.V. and Patil, A.N. 2014. Increase in growth and yield of pigeonpea with weed management. *Indian Journal of Weed Science* **46**(3): 264–266.

Hanumanthappa, D.C., Mudalagiriyappa, R., Veera, Kumar, G.N. and Padmanabha K. 2012. Effect of weed management practices on growth and yield of cowpea (*Vigna unguiculata* L.) under rainfed conditions. *Crop Research* **44** (1 & 2): 55-58.

http://agritech.tnau.ac.in/crop_protection/crop_prot_crop_insect_pul_bla%20and%20green.html

http://agritech.tnau.ac.in/crop_protection/crop_prot_crop_insectpest%20_cereals_maize.html.

http://agritech.tnau.ac.in/crop_protection/crop_prot_crop_insectpest%20_cereals_ragi.html

http://cropdisease.cropsci.illinois.edu/corn/Bacterialstalkrot.html

http://farmer.gov.in/imagedefault/ipm/Sesame.pdf

http://nsdl.niscair.res.in/jspui/bitstream/123456789/493/1/revised%20insect% 20pest%20 and%20their%20management.pdf

http://vasat.icrisat.org/?q=node/211

http://vikaspedia.in/agriculture/crop-production/integrated-pest-managment/ipm-for-pulses/ipm-strategies-for-blackgram-greengram/insect-mite-and-nematode-pests-management

http://vikaspedia.in/agriculture/crop-production/integrated-pest-managment/ipm-for-oilseeds/ipm-strategies-for-sesame/sesame-insect-pests-management

http://www.ncipm.org.in/NCIPMPDFs/IPM%20Packages%20DPPQ&S/Black Gram Green Gram. pdf

http://www.oisat.org/downloads/field_guide_sesame.pdf

https:// Agropedia, it is developed under the sponsorship of the ICAR, NAIP, 2011 - 2016 Department of Crop Science, North Carolina State University.

https://www.daf.qld.gov.au/plants/field-crops-and-pastures/broadacre-field-crops/maize/diseases

https://www.google.co.in/search?q=insect+pests+of+ragi &biw=1366&bih=662 &tbm=isch &tbo=u &source=univ &sa=X&ved=0ahUKEwj 6ifSh4vDQAhUEsY8KHQC4Dfc QsAQIOw &dpr= 1#imgrc=tXlJyUbRN9iMuM%3A

Komal, Singh, S.P. and Yadav, R.S. 2015. Effect of weed management on growth, yield and nutrient uptake of greengram. *Indian Journal of Weed Science* **47**(2): 206–210.

Kumar Anil, Kumar Jai, Puniya R., Mahajan Amit, Sharma Neetu and Stanzen Lobzang. 2015. Weed management in maize-based cropping system. *Indian Journal of Weed Science* **47**(3): 254–266.

Kumar, A., Saxena, A. and Singh, P.K. 2014. Chemical and mechanical weed management for increased yield of French bean. *Indian Journal of Weed Science* **46**(4): 350–352.

Marschner, H. 1995. Mineral Nutrition of Higher Plants. Academic Press, Orlando, Florida, USA

Mengel, K. and Kirkby, E.A. 1982. *Principles of Plant Nutrition.* International Potash Institute, Bern, Switzerland

Package of Practices for crops of Punjab Kharif 2013-14, Punjab Agriculture University Edited by Dr. H. S. Bajwa

Package of Practices for crops of Punjab Kharif 2014-15. Punjab Agriculture University. Edited by Dr. H.S. Bajwa.

Package of Practices for Kharif Crops. 2015. Sher-e-Kashmir University of Agriculture Sciences and Technology-Jammu, Directorate of Extension Education.

Panotra, N., Singh, O.P. and Kumar, A. 2012. Effect of chemical and mechanical weed management on yield of French bean–sorghum cropping system. *Indian Journal of Weed Science* **44**(3): 163–166.

Patel, K.R., Patel, B.D., Patel, R.B., Patel, V.J. and Darji, V.B. 2015. Bio-efficacy of herbicides against weeds in blackgram. *Indian Journal of Weed Science* **47**(1): 78–81.

Ram, S. and Verma, H.P. 2015. Weed growth, yield and quality of summer cowpea [*Vigna unguiculata* (L.) Walp.] as influenced by weed management. *Annals of Agricultural Research* **36** (3): 304-308.

Rana. M. C., Sharma, G.D., Sharma, A. and Rana, S.S. 2004. Effect of weed management and fertility levels on rajmash *(Phaseolus vulgaris)* and associated weeds under dry temperate high hills in Himachal Pradesh. *Indian Journal of Weed Science* **36**(3 &4): 227-230.

Rao, P.V., Reddy A.S. and Rao, Y.K. 2015. Effect of integrated weed management practices on growth and yield of pigeonpea [*Cajanus cajan* (L.) Millsp.] *International Journal of Plant, Animal and Environmental Sciences* **5**(3): 124-127.

Sharma, G.D., Sharma, J.J. and Sood, Sonia 2004. Evaluation of alachlor, metolachlor and pendimethalin for weed control in rajmash *(Phaseolus vulgaris* L.) in cold desert of North-Western Himalayas. *Indian Journal of Weed Science.* **36**(3 & 4): 287-289.

Sharma, J.C, Prakash, Chandra, Shivran, R.K and Narolia, R.S. 2014. Integrated weed management in pigeonpea [*Cajanus cajan* (L.) Millsp.] *IJAAS* **2**(1&2) 57-62.

Singh, G., Kaur H., Aggarwal, N. and Sharma, Poonam. 2015. Effect of herbicides on weeds growth and yield of greengram. *Indian Journal of Weed Science* **47**(1): 38–42.

Singh, V.P., Guru, S.K., Kumar, A., Banga, A. and Tripathi, N. 2012. Bioefficacy of tembotrione against mixed weed complex in maize. *Indian Journal of Weed Science* **44**(1): 1–5.

Srivastava, K.P. 1996. A Textbook of Applied Entomology (Vol. II). Kalyani Publishers, New Delhi, India.

Takkar, P.N. 1996. Micronutrient research and sustainable agricultural productivity in India. *Journal of the Indian Society of Soil Science.* **44**: 562-581.

Tisdale, S.L., Nelson, W.L., Beaton, J.D and Havlin, J.L. 1997. Soil Fertility and Fertilizers (Fifth edition). Second Indian Print, Prentice Hall of India Ltd, New Delhi.

Vasantharaj David, B. 2003. Elements of Economic Entomology. Popular Book Depot, Chennai, India

Yadav, K.S., Dixit, J.P. and Prajapati, B.L. 2015. Weed management effects on yield and economics of blackgram. *Indian Journal of Weed Science* **47**(2): 136–138.

About the Authors

Dr. Reena, completed her B.Sc. (Agriculture) from Banaras Hindu University, Varanasi, M.Sc. (Agril. Entomology) from University of Agricultural Sciences, Dharwad and Ph.D. (Entomology) degree from C.C.S. Haryana Agricultural University, Hisar. She qualified NET conducted by A.S.R.B., New Delhi and CSIR, New Delhi. She started her professional career as Assistant Professor-cum-Junior Scientist from July, 2004 at SKUAST-Jammu and has thirteen years of experience in Research, Extension work and teaching. She is also the recipient of Junior Research Fellowship (ICAR) during MSc. (Ag) degree program and Department of Science and Technology (DST), Government of India, Young Scientist Project Award (2009- 2012) under fast track scheme for young scientists. She is life member of several professional societies and has handled two externally funded project as PI and two as Co-PI. She has handled about ten University funded research projects as PI and Co-PI. She has delivered several expert lectures to the Department Agriculture personnel, farmers, pesticide dealers and other agriculture workers. She has published more than 25 research papers in journals of national and international repute. She has 10 book chapters, two research manuals and more than ten technical bulletins to her credit.

Dr Sonika Jamwal, working as Scientist Senior Scale (Plant Pathology) at Advanced Centre for Rainfed Agriculture, Dhiansar, SKUAST-Jammu has done her B.Sc. (Agriculture) from C.C.S University Meerut, M.Sc. (Plant Pathology) from SHIATS-Allahabad and Ph.D. in Plant Pathology from SHIATS from Allahabad.

She is specialized in biological control agents of soil borne fungus. She is working at SKUAST-Jammu since June 2007 and has ten years of experience. She is life member of several societies and she has handled two projects as P.I. and six as Co. P.I. She has delivered several expert lectures and has so many Radio and doordarshan talks. She has 20 research papers, 10 book chapters, 30 articles in magazines and newspapers, 15 pamphlets 2 booklets, 1 book and one manual on plant diseases to her credit.

Professor Anil Kumar graduated in Agriculture form the SKUAST-J&K, obtained his M.Sc. in Agronomy from HPKV-Palampur and Ph.D. from PAU-Ludhiana. He has served the state and SKUAST-J in different capacities including Agriculture Extension Officer, Assistant Professor Agronomy, programme coordinator of KVK and is presently working as Professor and PI, AICRP-Weed management in the Division of Agronomy, Sher-e- Kashmir University of Agricultural Science and Technology-Jammu. He has more than sixty research papers in national and international journals to his credit, two books and more than thirty chapters in various edited books. He was involved in 8 research projects taken up in the Division from time to time funded by various agencies like, NATP (ICAR), HTMM-I (ICAR), AICRP etc. as PI, life member of nine national and four international scientific societies, vice present of one and councillor of two national societies besides being on the panel of referees for various national and international journals. He is extensively involved in teaching UG, PG courses for the last fifteen years having guided a good number of M.Sc. and Ph.D students and has organized two summer schools sponsored by ICAR.

Dr. A. P. Singh is presently serving as Sr. Scientist (Agronomy) and Incharge, All India Coordinated Research Project on Dryland Agriculture (AICRPDA), Rakh Dhiansar, SKUAST-Jammu. He has earned more than 12 years of professional experience while serving in various capacities as Subject Matter Specialist, Associate Professor and Sr. Scientist. He has handled externally funded project on climate resilient agriculture in rainfed areas. He has 26 research papers published in peer reviewed journals of national and international repute. Apart from this, he has 13 popular scientific articles, 15 technical bulletins and other scientific publications such as book chapter etc to his credit.

Dr. Vikas Abrol, Soil Scientist received Ph.D degree from Sher-e-Kashmir University of Agricultural Sciences and Technology-Jammu and pursued post doctorate from Institute of Soil, Water and Environmental Sciences, Volcani Centre, Agricultural Research Organization, Israel.

He started his professional career as Assistant Professor/Jr. Scientist at Dryland Research Substation, Dhiansar and published research accomplishments in the journals of international repute like European Journal of Agronomy; European Journal of Soil Science, Journal of Soil and Sediments, Journal of the Science of Food and Agriculture, Agricultural

Mechanization in Asia, Africa and Latin America etc with high impact factor and edited two international books "Crop Production Technologies" and "Resource Management for Sustainable Agriculture". He was involved in externally funded projects as Principal Investigator and Co- Principal Investigator and also served as reviewer of research articles in international journals like Soil Science Society of America Journal, Land Research and Development, Agronomy for Sustainable Development, Soil Science, Indian Journal of Agricultural Sciences, Indian Journal of Dryland Agricultural Research and Development, Indian Journal of Soil Conservation, Journal of Experimental Biology and Agricultural Sciences and International Journal of Agriculture Sciences etc. His research interests entail using biochar for offsetting climate change by soil carbon aggradations, runoff quality and soil erosion control, soil quality evaluation, soil and water pollution.

Dr. Jai Kumar has been presently working as Jr. Scientist, Sr. Scale (Agronomy with Specialization in Field crops, Weed Management in Rainfed Agriculture) at Advance Centre for Rainfed Agriculture, Rakh Dhiansar. His research interests entail Field crops under dryland conditions, Alternate Land Use System, Inter Cropping and soil erosion mitigation. He has worked as Co-PI in many research projects (AICRPDA, NICRA, NABARD) and is involved in research, teaching and extension since 2006. He has published 20 research papers in Journals of national and international repute.

Dr. R. Puniya graduated in Agriculture form the RAU, Bikaner, obtained his M.Sc. in Agronomy and Ph.D. from GBPUAT, Pantnagar. He also served as post doctoral fellow in conservation agriculture at IARI, New Delhi. He is presently working as Jr. Scientist in AICRP-Weed Management, Division of Agronomy, Sher-e-Kashmir University of Agricultural Science and Technology-Jammu. He has about more than fifteen research papers in national and international journals to his credit, two book, and four book chapters in various edited books. He was involved in 3 research projects taken up in the Division of Agronomy from time to time funded by various agencies. He is also life member of three national scientific societies. He is extensively involved in teaching UG, PG courses for the last five years having guided two M.Sc. students.